CLASS NOTES

SYGN 202 – Engineering Materials Systems

Introduction to Materials Science

From Atom to Component*

by

Hans-Joachim Kleebe

Full Professor
Colorado School of Mines
Metallurgical and Materials Engineering Department
Golden, CO 80401, USA

* **Cover Page**
High-resolution TEM image of SiC showing the atomic structure (top)
TEM micrograph revealing the overall microstructure of SiC (middle)
Optical image of SiC components (bottom; courtesy CoorsTek)
Note that all images were taken from the same material; however, at different magnifications

PEARSON
Custom Publishing

Printed in the United States of America

10 9 8 7 6 5 4 3 2

ISBN 0-536-84709-6

2004360416

AG

Please visit our web site at *www.pearsoncustom.com*

PEARSON CUSTOM PUBLISHING
75 Arlington Street, Suite 300, Boston, MA 02116
A Pearson Education Company

Table of Contents

Preface

This booklet is intended for freshmen and sophomore students attending the SYGN 202 Materials Engineering Systems class at the Colorado School of Mines.
The **class notes** cover the material taught in the classroom and are designed to help the student follow the progress of the class more easily. However, it should be noted that the text provided is not a stand-alone textbook; hence, *attending class is mandatory*.

These class notes are based on two main textbooks: (i) J.F. Shackelford, "Introduction to Materials Science for Engineers" [a] and (ii) W.F. Smith, "Foundation of Materials Science and Engineering" [b].

The main intention of creating a shortened version of these excellent textbooks was a response to the students. They suggested that it would be helpful to follow the class ("since we cover a lot of ground") when a text is at hand that only presents the material covered in the classroom. The result of the students' request is this brief introduction to materials science.

However, both textbooks mentioned above are strongly recommended to the students for further reading, since they cover more material than presented in SYGN 202 and they also present many example problems which are very helpful to understand the basic concepts of material science and engineering.

I would like to thank two colleagues of mine, Prof. Ivar Reimanis and Prof. John Speer, who were very supportive while putting the material for these class notes together.

Moreover, I would like to express my thanks to those students, who were willing to read the manuscript draft and add important constructive criticism. In particular, I would like to thank Elanor Steffee and Stacy Carrera, who were willing spend part of their spare time to suggest helpful corrections. Tim Casias was very supportive in preparing drafts of some of the figures shown in the text and cross-reading the manuscript carefully.

Last but not least, I want to express my sincere thanks to my family, my wife Sylvia, and our sons André and Simon, for their support during the time these class notes were edited; without their loving understanding I would not have succeeded.

Hans-Joachim Kleebe

---

[a] J.F. Shackelford, "Introduction to Materials Science for Engineers," 5th Edition, Prentice Hall, New Jersey, 2000.

[b] W.F. Smith, "Foundations of Materials Science and Engineering," 2nd Edition, McGraw Hill, New York, 1993.

1 Introduction

This class gives a brief overview of the different material classes and introduces the fundamental concepts underlying material behavior and properties. In particular, the following topics will be addressed:

- Atomic bonding, crystal structures and intrinsic defects
- Phase diagrams
- Material properties (mechanical, thermal, optical, electrical, magnetic)
- Differences between the five material classes
- Origin of their specific characteristics
- Processing techniques to control micro-structure and material properties to meet application needs.

1.1 Materials

We live in a world of various materials, which largely define our social relationships as well as the economic quality of our daily life (e.g., car, refrigerator, dishwasher, television, mobile phone, etc.).

Material possessions were also important for our ancestors. Most important for them were **tools & weapons**. Time periods were named after the material that was predominantly used for these items at that time.

- Stone age — goes back to about 2.5 million years
- Copper age — 4000-3000 B.C. (in Europe)
- Bronze age — 2000-1000 B.C.[1]
- Iron age — 1000-500 B.C.

[1] **foundation of metallurgy**, alloys composed of Cu+Sn

Note that pottery and glass artifacts have been traced back to about 4000 B.C.

Today's world includes many **polymeric materials** (bottles, containers, toys) and **silicon-based devices** (electronic industry), due to their pervasive impact on our lives. Materials are very important in every technical discipline ranging from steel making to aerospace applications.

There are various ways of **classifying materials**:

- Crystal structure and atomic bonding
- Chemical composition (Al, AlN, Al_2O_3)
- Processing (cold working, rolling, pressing, casting, stamping, sintering, hot pressing, HIPing)
- Shape (I-beam)
- Application (bike frame, ski, fender)
- Properties (strong, brittle, lightweight, magnetic, conducting, insulating).

1.2 General Characteristics of the Different Material Classes

1.2.1 Metals

Metals are strong and ductile (which means they can readily be formed into complex shapes), are very good thermal and electrical conductors and show metallic luster (cobalt, copper, silver, gold, platinum).

Alloys: are metals composed of more than one element; however, no separate compounds or phases are typically formed within an alloy.

1.2.2 Ceramics

Ceramics are materials composed of a metal cation and a non-metallic anion. One example is the oxidation product of Al: $Al \rightarrow Al_2O_3$ (alumina). Ceramics are usually hard and brittle, have strong bonds and corresponding high melting points, are non-conducting and chemically stable.

Glasses: belong to the family of ceramics but do not reveal a long-range ordered crystal structure. Their atoms are stacked in an irregular, random pattern, which is termed an **amorphous** structure.

Note that glasses do not show a distinct melting temperature. With increasing temperature, glass viscosity continuously decreases until the glass melt formed. The onset temperature of this behavior is called transition temperature.

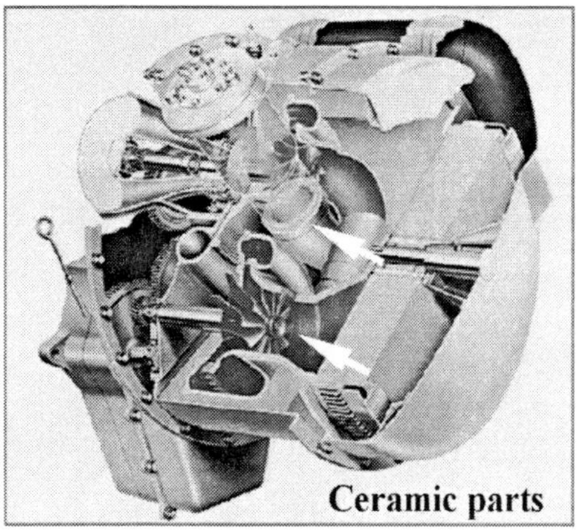

Fig. 1.1 Automobile engine equipped with various ceramic parts (arrows).

1.2.3 Polymers

Polymers are a class of materials that typically contain long chains of low weight elements (C, H, N, O, F). They are characterized by low electrical conductivity, usually low strength, and low melting points (decomposition). Polymers are lightweight with high chemical reactivity (UV sensitive), but can easily be processed into complex shapes.

1.2.4 Composites

Composites are a combination of different materials, which provide better properties than either of the materials alone (strength, stiffness, fracture toughness).
Fiberglass = glass fibers within a polymer matrix (strong vs. ductile).
Concrete = aggregate composite of pebbles and sand within a silicate cement matrix.
Wood = natural fiber-reinforced composite.
Bone = composite of hydroxylapatite (HAp, $Ca_{10}(OH)_2(PO_4)_6$) and collagen.

Fig. 1.2 Left is shown wood, a natural fiber-reinforced composite, while to the right concrete is shown, a synthetic aggregate composite.

1.2.5 Semiconductors

Semiconductors are neither good conductors nor highly insulating. Elements like Si and Ge are semiconducting, as are nearby III-V compounds like GaAs and InP. The

elements Si, Ge, and Sn can be seen as the boundary between metallic and non-metallic elements in the periodic table.
Semiconductors are used in microcircuits as rectifiers (GaAs), solar cells (CdS, CdInSe), and phosphors in color television screens (doped-ZnO), among many other important applications.

Example 1.1

The next generation of commercial aircrafts is currently under development. The ultimate goal for commercial aircrafts is to reach a speed of Mach 3. There is a strong competition in this field between U.S., Japan and Europe. To achieve that goal, the required engine temperature should be on the order of 1500 °C in a combustion environment. As a consequence, **new designs** have to be developed. What materials would be potential candidates? *Metallic alloy composites or ceramic matrix composites.*

1.3 Overall Concept of Class

The class is structured in a way to first present a basic understanding about atomic bonding and crystal structures including crystal defects. Based on the understanding of structures and phases, phase diagrams can be used as a tool to understand why different microstructures evolve. Basic mechanical concepts will be presented and applied to the different material classes, which will lead to an understanding of material properties and material behavior under experimental or service conditions. During the description of the different materials classes, the correlation between bonding/structure and properties will be emphasized.

The aim of this class is that the students generate a "feeling for materials" and intuitively know why, for example, some materials fail while others last (corrosion, fatigue, thermal shock resistance).

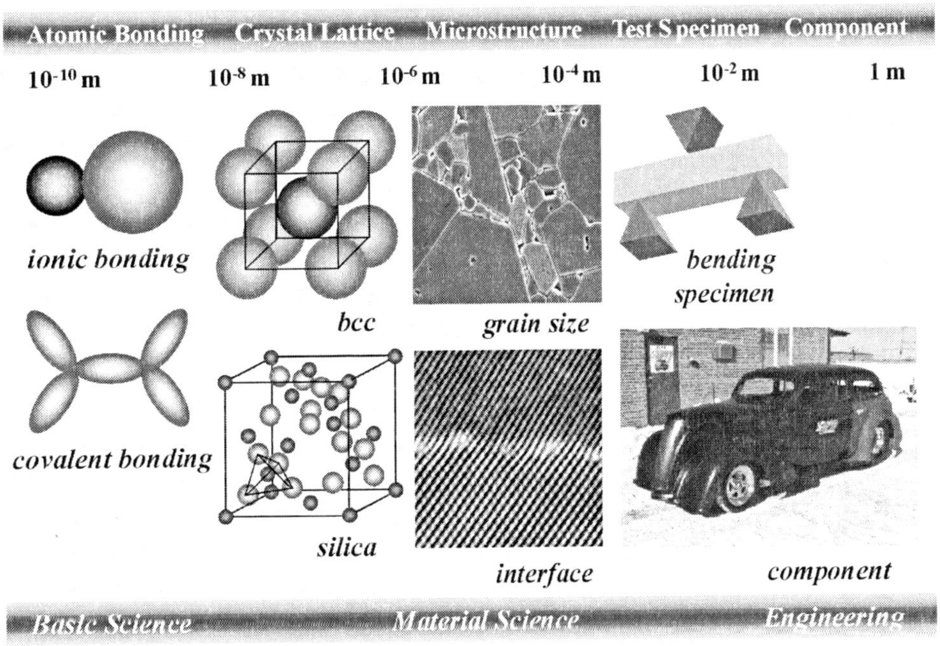

Fig. 1.3 From Atom to Component.

2 Atomic Bonding

2.1 Atomic Structure

The type of bonding within a material controls the intrinsic **structure** of solids and, therefore, dominates their **properties**. In order to understand atomic bonding, we must appreciate the structure within individual atoms.

For the purpose of general understanding of crystal structures and material properties, the rather simplified representation of the atomic structure via the **shell model** is used here. Figure 2.1 shows the planetary model with the *positively* charged **nucleus** at the center and the *negatively* charged **electrons** orbiting about it.

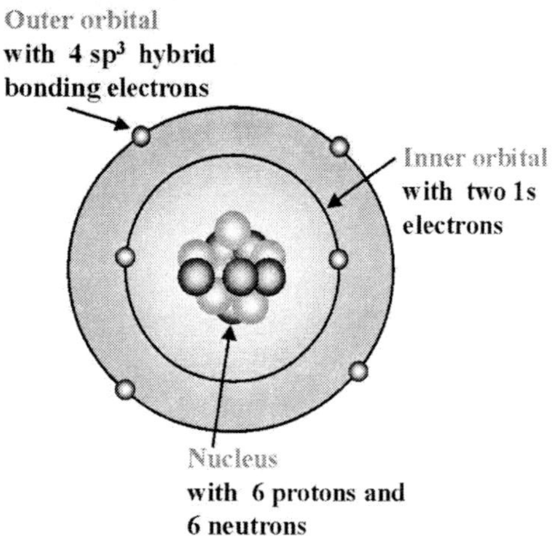

Fig. 2.1 Simplified shell model of a single carbon atom.

The **nucleus** is composed of **protons** and **neutrons** with masses of **$1.66 x 10^{-24}$ g** and contains nearly all the mass of the atom, while the electron has a much lower mass of **$0.911 x 10^{-27}$ g** and a negative charge of **$-0.16 x 10^{-18}$ C** (Coulomb)[2].

[2] **Charles Augustin de Coulomb** (1736-1806), French physicist.

The **atomic number**, **Z**, equals the number of protons (positively charged particles) within the nucleus. *Z* also equals the number of electrons around the nucleus when the atom is in a neutral state; see also the **periodic table.**

The concept of the periodic table was first published by **D.I. Mendelejev** [3] in 1869, based on the principle of increasing atomic weight.
The **atomic weight** or mass of an element is the mass of one mole of that element (in grams), where one mole is the **Avagadro's** [4] **number** of atoms, **$N_A = 6.023 x 10^{23}$** atoms.

Carbon, **^{12}C** (6 protons and 6 neutrons) has an atomic mass of 12.000 *g* and is used as reference mass. One **atomic mass unit**, *u*, is defined as exactly 1/12 of the mass of a ^{12}C atom, which could simply be calculated by taking Avogadro's number and multiplying it with the mass and number of protons and neutrons.

$6.023 x 10^{23} x 1.66 x 10^{-24}$g x 12 = 11.998 g

The small difference is related to the fact that the electron mass was not considered here. For copper, **Cu**, with Z=29 the atomic mass is **63.54**. ***Note*** that copper has two principal isotopes Cu^{63} and Cu^{65}. Therefore, the atomic masses are not integers.

[3] **Dimitri Ivanovitch Mendelejev** (1834-1907), Russian physicist.

[4] **Amadeo Avogadro** (1776-1856) was an Italian physicist, who coined the term molecule. His hypothesis that all gases at a given temperature and pressure contain the same number of molecules was only generally acknowledged after his death.

Isotopes of one element have the same atomic number but different atomic masses, due to different number of neutrons: ^{12}**C** vs. ^{13}**C**. The average atomic mass of carbon is 12.01 because of these isotopes.

Example 2.1

(a) how much does one Cu atom weigh?

The atomic mass of Cu = 63.54. This means that one mole Cu weighs 63.54 g = 63.54 g/mol.

$$\frac{63.54\, g / mol\, Cu}{6.023 \cdot 10^{23}\, atoms / mol} = \frac{xg\, Cu}{1\, atom}$$

=> mass of one Cu atom = 1.05x10^{-22} g

(b) how many Cu atoms are in 1g?

$$\frac{6.023 \cdot 10^{23}\, atoms / mol}{63.54\, g / mol\, Cu} = \frac{x\, atoms\ Cu}{1\, g\, Cu}$$

=> number of Cu atoms = 9.47x10^{21} atoms.

While chemical identification is done relative to the nucleus (number of protons *and* neutrons), **atomic bonding involves** the **electrons** (electron orbitals). The mass of an electron, 0.911x10^{-27} g, is a negligible contribution to the total mass; however, the electron has a negative charge of –0.16x10^{-18} C, which compensates for the positive charge of the proton +0.16x10^{-18} C. As a result, atoms/elements reveal **charge neutrality**.

Each **orbital radius** (see shell model in Figure 2.1) is characterized by an **energy level**, which corresponds to a fixed **binding energy** or energy needed to separate the electron completely from the atom, as shown in the energy-level diagram of ^{12}C (Figure 2.2).

Quantum physics predicts that there are discrete energy levels (with forbidden energy levels in between). ***Note*** that the *sign convention* refers to a negative energy as an attractive force.

^{12}**C**: **1s^2 2s^2 2p^2** equals six electrons, the outer four electrons redistribute and create a new and characteristic configuration, the so-called **sp^3 hybridization** (see also covalent bonding section and Figure 2.12).

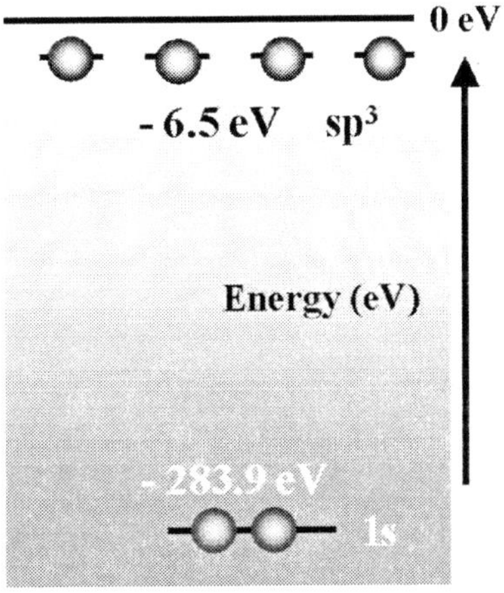

Fig. 2.2 Energy-level diagram of ^{12}C.

2.2 Types of Atomic and Molecular Bonds

Bonding of adjacent atoms is essentially an **electronic process** since it involves outer valance electrons. There are two major types of bonding:

- Strong **primary bonds** are formed when outer orbital electrons are transferred or shared between atoms. Primary bonds *always* involve **electron transfer** or **sharing electrons.**

	Binding Energy
▪ **Ionic**	500-1000 kJ/mol
▪ **Covalent**	400-750 kJ/mol
▪ **Metallic**	70-800 kJ/mol

- Weaker **secondary bonds** (van der Waals bonding) result from a more subtle attraction between positive and negative charges (**dipoles**) with no actual transfer or sharing of electrons.

The major source of **cohesion** in a given engineering material is due to one or more of the three primary bonds; their bond energy ranges between 70-1000 kJ/mol. All primary bonds are rather strong bonds resulting in materials with relatively high melting or dissociation temperatures.

The general **driving force** for bonding is to lower the overall energy of the bonding electrons.

2.2.1 Ionic Bonding

The ionic bond is a result of **electron transfer** from one atom to another. Electron transfer exists when the **electronegativity** difference between elements is rather large, as in case of NaCl or CsCl.
Electronegativity can be described as the attractive force between the outer valance electrons and the nucleus of one atom.

Na^0 gives up its outer $3s^1$ electron and **transfers** it to a half filled 3p orbital of Cl atom (filled p-shell) resulting in alternating Na^+ and Cl^- ions. This transfer is favored because it produces a **more stable electronic configuration**.

The positively charged cation, Na^+, and the negatively charged anion, Cl^-, both have a filled outer orbital electron shell due to the transfer of one electron (Figure 2.3), which results in the **noble-gas configuration** ($Na^+ \leftrightarrow Ne^0$, $Cl^- \leftrightarrow Ar^0$).

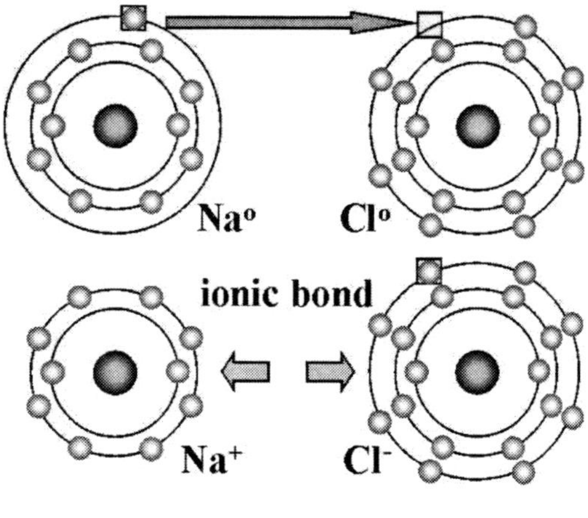

Fig. 2.3 Schematic of the electron transfer between Na and Cl.

The **ionic bond** is **non-directional**; packing occurs with no preferred orientation. Any Na^+-ion will attract any adjacent Cl^--ion equally in all directions, as shown in Figure 2.4.

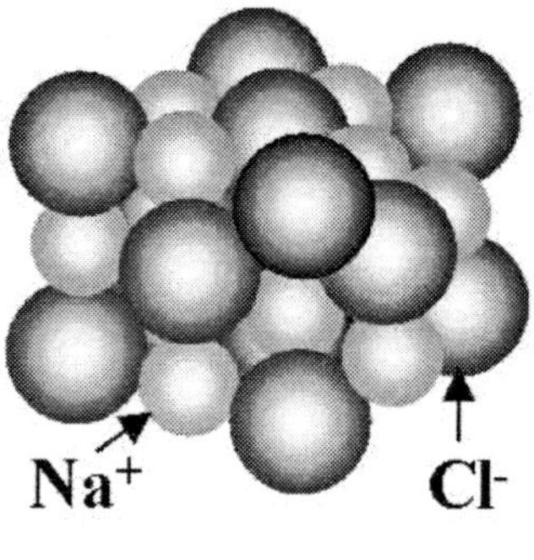

Fig. 2.4 NaCl structure with Na^+ and Cl^- ions equally attracting each other in all directions.

The ionic bond is a result of the **Coulombic attraction** between oppositely charged ions.

Coulomb force: $$F_c = \frac{-K}{a^2} \qquad (2.1)$$

With a being the separation distance between the centers of the ions and K equals:

$$K = k_0 \left(Z_1 q\right)\left(Z_2 q\right) \qquad (2.2)$$

Here, k_0 is a proportionality constant, $9.0x10^9$ Vm/C, with Z_i is the valence or, in other words, the net charge of the atom (Na^+ = +1) and q is the electron charge of $-0.16x10^{-18}$ C.

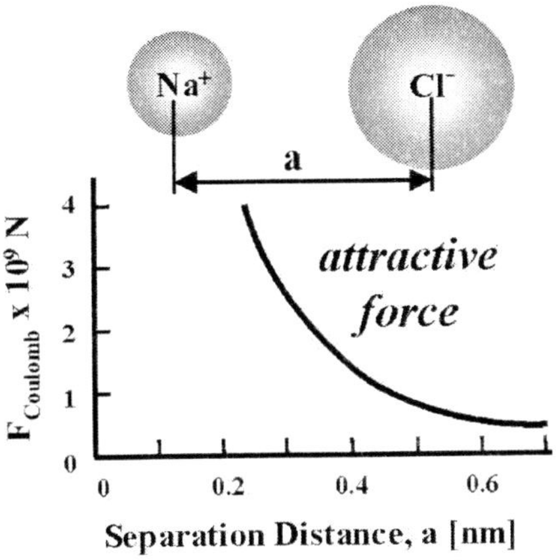

Fig. 2.5 Coulomb attraction vs. separation distance.

The attractive Coulomb force, F_c, increases, when the ions get closer together, as shown in Figure 2.5. Ideally, the separation distance, a, should be zero, however, a **repulsive force**, F_R, counteracts the attractive Coulomb force, owing to **overlap** of the electron clouds and repulsion of the two positively charged nuclei (Figure 2.6).

Repulsive force: $F_R = \lambda e^{-a/\rho}$ (2.3)

λ and ρ are experimentally determined constants for a given ion pair.

Bonding force, F_B, is the **net force** of attraction and repulsion as a function of the separation distance, a (Figure 2.7).

Equilibrium bond length, a_0, exists at

$$F_C + F_R = 0 \qquad (2.4)$$

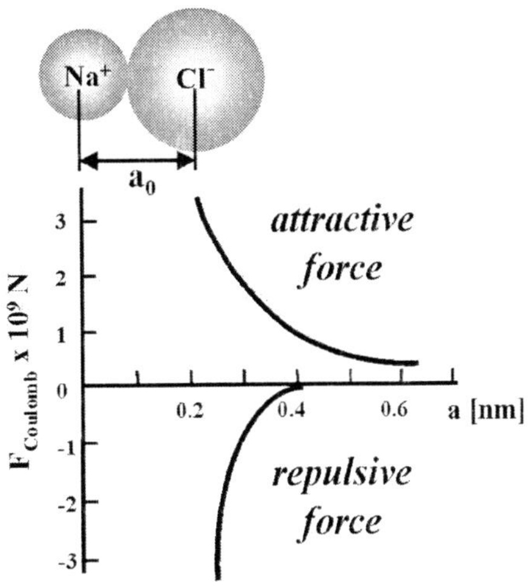

Fig. 2.6 Coulomb attraction and repulsion vs. separation distance.

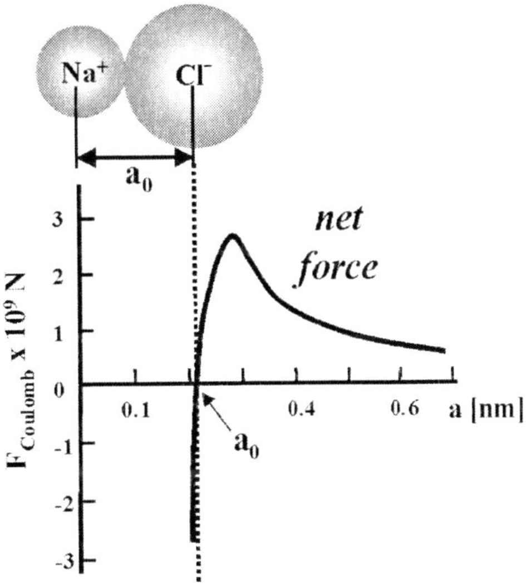

Fig. 2.7 Net force of attraction and repulsion as a function of the separation distance.

Since a_0 is an **equilibrium** bond length, it is a sum of two ionic radii: $a_0 = r_{Na^+} + R_{Cl^-}$. To change a_0, **externally applied forces** are required (compressive or tensile load), which has implications on the mechanical behavior of solids (elastic modulus).

The **bonding energy**, E_B, is related to the bonding force by the expression:

$$F_B = \frac{dE_B}{da} \quad or \quad E_B = \int F da \qquad (2.5)$$

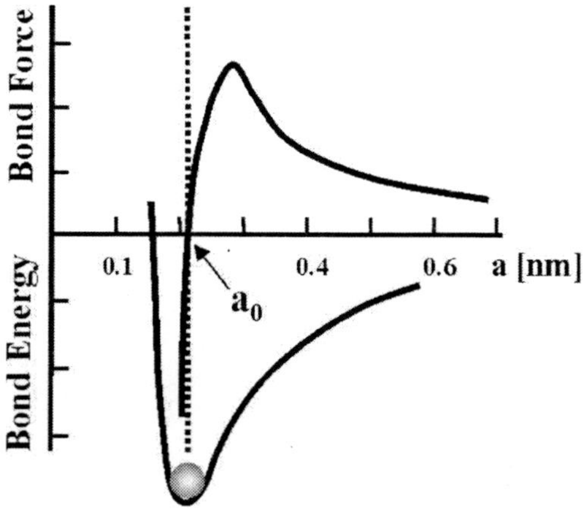

Fig. 2.8 Bonding energy and bonding force vs. separation distance. The derivative of the bond energy is the bonding force.

Hence, the net bonding force curve is the **derivative** of the **bonding energy** curve. The stable ion position (Figure 2.8) corresponds to an **energy minimum**:

$$F_B = \left(\frac{dE_B}{da}\right)_{a=a_0} \qquad (2.6)$$

Ionization involves a change in ionic radius: $Na^0 > Na^+$; $Cl^0 < Cl^-$. The ionic radius usually determines how all these atoms are packed in a structure, which leads to the **coordination number**, CN, that is the number of adjacent ions or atoms surrounding a given ion or atom.

2.2.1.1 Coordination Number

In NaCl (rocksalt), six Na^+ ions surround one Cl^- ion (and vice versa). Hence, the corresponding CN = 6; there are six nearest neighbors. In **ionic compounds**, the CN of the **smaller** ion can be predicted in a **systematic way** by considering the larger ion (of opposite charge) that can be in contact with the smaller one (Figure 2.9).

CN depends directly on the **relative sizes** of the different ions and is characterized by the **radius ratio, r/R**, which can be calculated as follows:

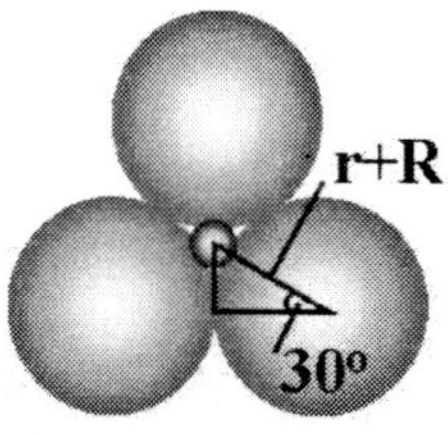

$$\cos 30^o = 0.866 = \frac{R}{r+R} \Rightarrow \frac{r}{R} = 0.155$$

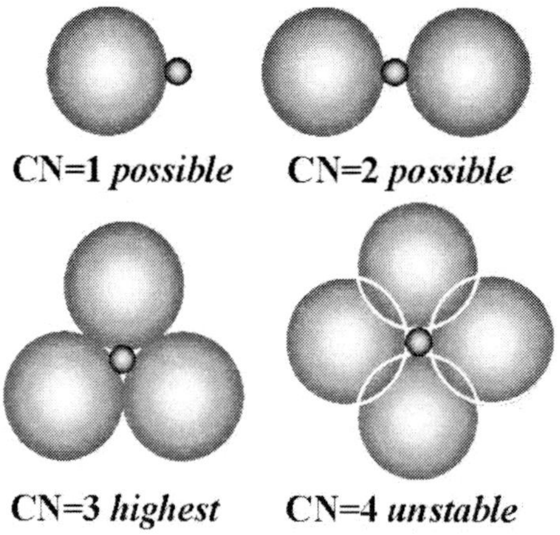

Fig. 2.9 Possible coordination numbers, CN, for r/R=0.155. *Note* that CN=4 is unstable (see also Table 2.1).

In this example, the **CN=4** is rather **unstable**, because the charges of the four outer ions overlap. On the other hand, CN=3 is stable up to a r/R=0.225 (where 4-fold coordination becomes possible).

The minimum radius ratio that can produce a coordination of **CN = 3** is **0.155** (Table 2.1).

Tab. 2.1 Correlation between radius ratio and coordination number.

Coordination Number	Radius Ratio
CN = 2	$0.000 \leq r/R \leq 0.155$
CN = 3	$0.155 \leq r/R \leq 0.225$
CN = 4	$0.225 \leq r/R \leq 0.414$
CN = 6	$0.414 \leq r/R \leq 0.732$
CN = 8	$0.732 \leq r/R \leq 1.000$
CN = 12	1.000

<u>Summary of ionic bonding</u>

- full electron transfer
- non-directional bond
- coordination number, CN, controlled by ionic radii
- atomic and ionic radii differ (gain or lose electron charge)
- pronounced difference in electro-negativity.

2.2.2 Covalent Bonding

While the ionic bonding is non-directional, the covalent bond is **highly directional.** The name *"covalent"* derives from **cooperative sharing** of valence electrons between two adjacent atoms. There is only a ***small* electronegativity difference** (shared outer s and p electrons).

Valence electrons are those outer orbital electrons that take part in the bonding process. There are different ways to illustrate the bond between shared electrons, as shown in Figure 2.10 for Cl_2-gas.

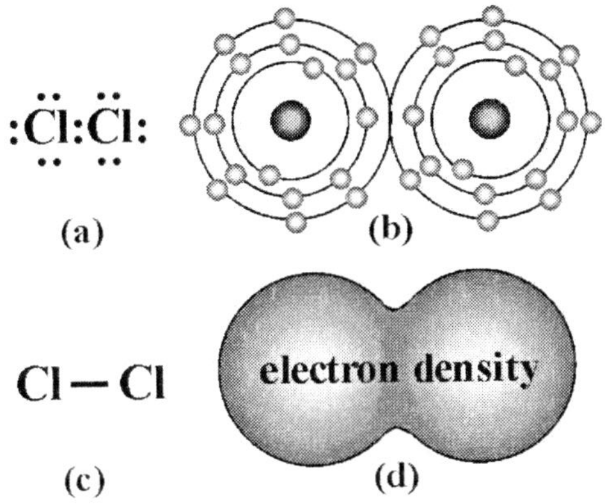

Fig. 2.10 Different ways to represent the covalent bonding (shared electrons) of a Cl_2-gas molecule.

Diamond with its exceptionally high hardness and a melting point, T_m, higher than 3550 °C, has covalent bonding between each adjacent pair of C atoms.

Note that covalent bonding can produce coordination numbers *substantially smaller* than predicted by the r/R-ratio of ionic bonding.

For diamond, the r/R-ratio is 1.0, but **CN = 4** rather than **12**, as predicted by the ionic model.

Carbon has an electron configuration of $\mathbf{1s^2 2s^2 2p^2}$ in its ground state; however, in this case, CN is determined by the characteristic **sp^3-hybridization**, in which the 4 outer shell electrons of the C-atom are shared with adjacent atoms in equally spaced directions (Figures 2.11 and 2.12).

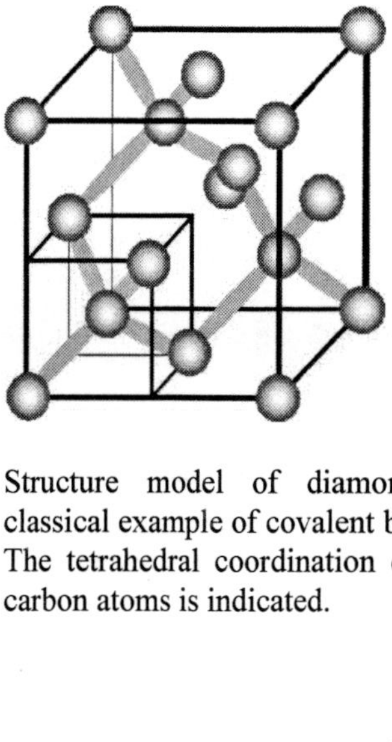

Fig. 2.11 Structure model of diamond; the classical example of covalent bonding. The tetrahedral coordination of three carbon atoms is indicated.

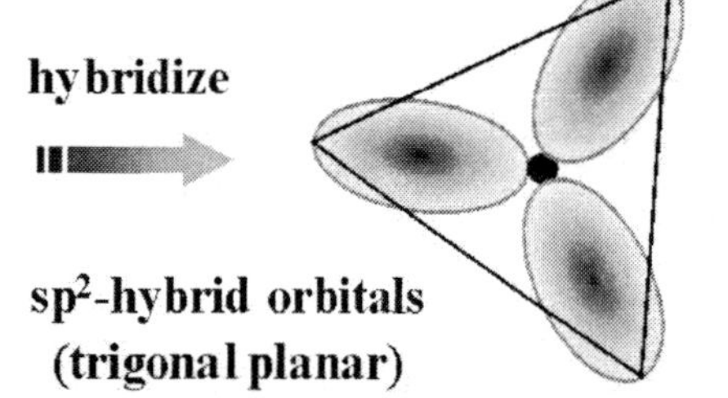

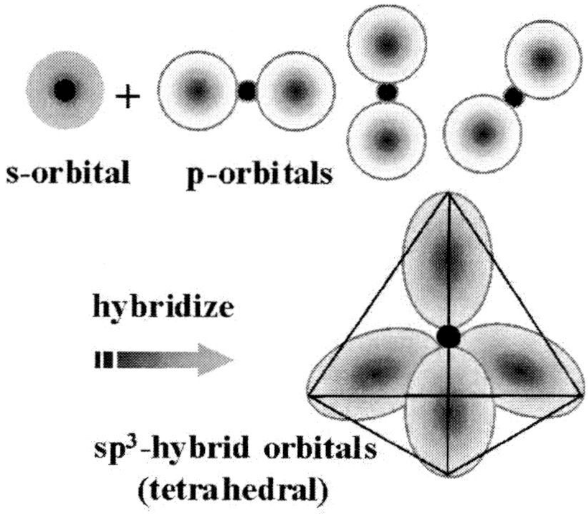

Fig. 2.12 Schematic representation of a sp^{2-} and sp^3-hybrid orbital. *Note* that various other configurations are also possible.

When bringing C atoms closer together, **sp^3 hybrid orbitals** are formed. There are four sp^3 hybrid orbitals created instead of individual p- and s-orbitals. Now four covalent bonds can be formed, each of equal strength (Figure 2.12). As a consequence, diamond is **extremely hard** and has a **high melting temperature** of about 3550 °C (it also has a high bond strength; compare Table 2.2).

The **SiO_4^{4-} tetrahedron** is the main building unit for many naturally occurring silicates like feldspar (a major component of granite). The **r/R-ratio** of the silica tetrahedron equals 0.039nm/0.132nm = **0.295**, which agrees well with the ionic model given a **CN = 4**. This is due to the fact that the Si-O bond is 50% ionic plus 50% covalent; giving a **mixed bonding character** (a common property of many natural and synthetic compounds).

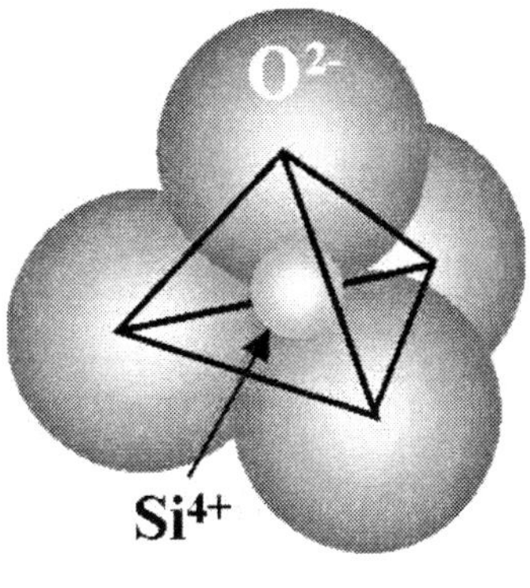

Fig. 2.13 Structure model of the $(SiO_4)^{4-}$ tetrahedron.

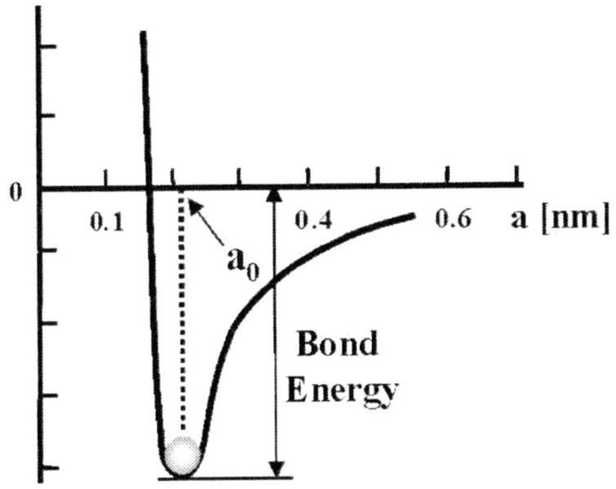

Fig. 2.14 Bonding energy vs. separation distance. The derivative of the bond energy is the bonding force.

Note that most bonding types are of mixed character and do not represent a pure bonding type. Commonly, bonding is described as of more or less ionic (covalent, metallic) character.

Tab. 2.2 Bond energy and bond length of different bonds.

Bond	Bond Energy	Bond Length
C-C	370 kJ/mol	0.154 nm
C=C	680 kJ/mol	0.130 nm
C≡C	890 kJ/mol	0.120 nm
Si-O	375 kJ/mol	0.160 nm
O-H	500 kJ/mol	0.100 nm

The bonding force and bonding energy curves for covalent bonding look very similar to those shown for ionic bonding (Figure 2.7). The same is true of **metallic** and secondary bonding, which will be discussed in the following sections.

2.2.3 Metallic Bonding

Metallic bonding involves **electron sharing** and is **non-directional**. In this case, the valence electrons are **"delocalized"**, that is, they have an equal probability of being associated with any of a large number of adjacent metal atoms.

In typical metals this delocalization is associated with the entire material, leading to an **"electron cloud"**, or **"free electron gas"** (see Figure 2.15). This highly mobile electron gas is the basis for the high electrical and thermal conductivity of metals.

As with ionic bonding, **bond angles** and **CN** of metallic compounds are determined primarily by **efficient packing** of atoms, therefore, CN tends to be as high as **8** or **12**.

Note the **fewer valence electrons** are in a metal, the more 'metallic' bonding is achieved, since electrons can move around more freely. Consequently, alkali metals (Li, Na, K, Rb, Cs, Fr) have the most metallic bonding character (only one valence electron) and a much **higher electrical conductivity**.

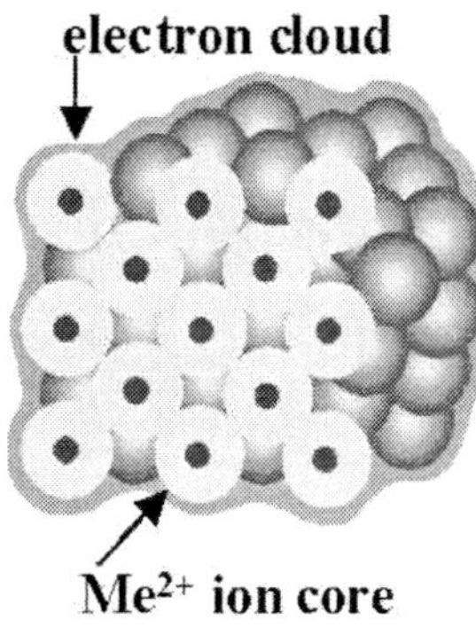

Fig. 2.15 Schematic of metallic bonding with metal core ions immersed in the free electron gas.

Metal core atoms can also **easily slide** past one another. As a result, metals show a rather **low yield stress** (stress level above which permanent deformation is initiated) in comparison to covalently bonded materials such as diamond.

2.2.4 Secondary Bonding

As we learned earlier, the major source of cohesion in a given engineering material is one or more of the three primary bonds; with bond energies ranging between 70-1000 kJ/mol. However, it is also possible to obtain bonding with substantially **smaller bond energies** and ***without* electron transfer** or electron **sharing**. This is known as secondary bonding or **van der Waals** [5] bonding, which involve electric dipoles. The overall **driving force** (incentive for bonding) is the **attraction** between different **electric dipoles**. Similar to ionic bonds, dipoles interact with one another through electrostatic or Coulomb forces (compare equations 2.1 and 2.2 in ionic bonding section).

This electrostatic attraction depends on an **asymmetrical distribution** of positive and negative charges within each atom or molecule. Such charge asymmetries are referred to as **dipoles.**

We have to distinguish between **two types** of secondary bonding:

- **temporary** (induced)
- **permanent**.

When for example two **Ar** atoms approach each other (Ar is a noble gas with filled outer orbital electron shell), the negative charge is slightly drawn towards the positive nucleus on the adjacent atom. This slight **charge distortion** occurs *simultaneously* in both atoms. The result is an **induced dipole** with a rather weak bond energy of about **0.99 kJ/mol.**

On the other hand, because of the *directional nature* of the covalent **O-H** bonds in **water**, the two H atoms become positive centers while the O atom become a negative center (Figure 2.16). Such a **polar** molecule is characterized by a **permanent** separation of the charges, which gives a larger dipole moment and hence also a greater bond energy of about **21.0 kJ/mol.** ***Note*** that bonding between polymeric chains is of the same type.

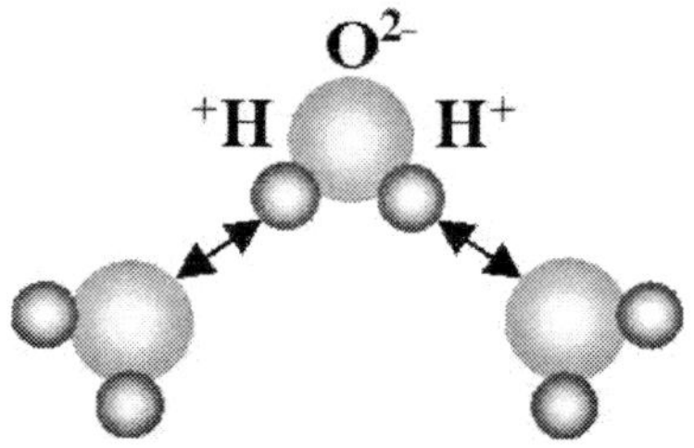

Fig. 2.16 Schematic representation of water with permanent dipole attraction (hydrogen bonds).

[5] **Johannes Diderik van der Waals** (1837-1923), Dutch physicist.

It is interesting to note that the expansion of water upon freezing is a result of a repeated **alignment** of **H_2O molecules** which leads to a relatively open structure. When melted, the H_2O molecules pack much more randomly and fit a smaller total volume (higher density).

Note that a common description of the bonding energy curve for secondary bonding is the so-called **6-12 potential**:

$$E = \frac{K_A}{a^6} + \frac{K_R}{a^{12}} \qquad (2.7)$$

where K_A and K_R are constants for attraction and repulsion, respectively, with values of $K_A = 10.37 \times 10^{-78}$ Jm^6 and $K_R = 16.16 \times 10^{-135}$ Jm^{12}.

2.3 Bonding Classification

An instructive presentation of relative bond energies of the various bond types is obtained by comparing respective melting points. The **melting point** of a solid is the temperature to which a material must be subjected in order to provide sufficient thermal energy to **break** its **cohesive bonds** (Table 2.3).

Note that when referring to the melting point of a polymer, there is no precise melting temperature. The polymer loses its structural rigidity over a wider temperature range since the weak secondary bonds between the polymer chains start to break first. As there are numerous polymer chain lengths, the decomposition of the polymer occurs over a range of temperatures.

Tab. 2.3 Bonding type vs. melting temperature of different materials.

Material	Bonding Type	Melting Point [°C]
Ar	**secondary** ***induced dipole***	**- 189**
H_2O	**secondary** ***permanent dipole***	**0**
C_2H_4	**cov. + second.**	**~120**
NaCl	**ionic**	**801**
Cu	**metallic**	**1084.87**
$C_{Diamond}$	**covalent**	**~3550**

Note that most natural compounds such as quartz, feldspar and mica (all of which compose the granite found in the Rocky Mountains) as well as most engineering materials typically do not reveal only one type of pure bonding, but rather a mixture of different bond types.

2.4 Mixed Bonding

Similar to composites where one can combine all kinds of materials with one another (ceramic fibers within a metal matrix or ferromagnetic particles in a polymer matrix), different combinations of bond characters are also possible:

- **ionic-covalent**
- **metallic-covalent**
- **metallic-ionic**
- **covalent-secondary**

For example, semiconductor compounds (GaAs, InP, CdTe) have mixed covalent-ionic bonding (see Table 2.4 and Figure 2.17). The percentage of the particular mixed-bonding character depends on the **electronegativity difference**.

Tab. 2.4 Mixed bonding character of different materials.

Material	Bonding Type	Example
Metals	**metallic-covalent**	**Cu, alloys**
Ceramics	**ionic-covalent**	**Si_3N_4, SiO_2**
Polymers	**covalent-second.**	**C_2H_4**
Semicon-ductors	**covalent-ionic**	**Si, Ge, InP**

Transition metals are largely mixed **metallic-covalent**; as you go to left side of the periodic table, the bonding character gets more metallic, while the right side shows more covalent type of bonding. The same holds for going down a column, e.g., C = covalent, Si and Ge = mixed, Sn and Pb = metallic.

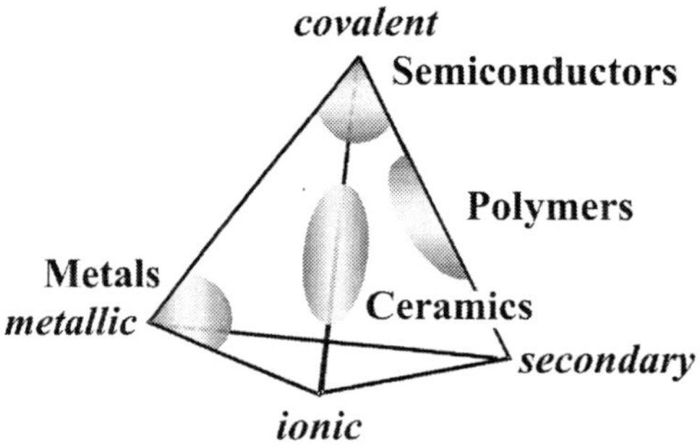

Fig. 2.17 Bonding tetrahedron indicating the mixed bonding character of the different material classes.

Summary of bond types

While the chemical behavior of each atom is determined by its nucleus (number of protons and neutrons), the nature of atomic bonding is determined by the behavior of the electrons that orbit the nucleus. Cohesion of solids is given by the three strong primary bonds and the two weak secondary bonds:

- **ionic** - ***non-directional, full electron transfer***, which creates two ions that are attracted by Coulomb interaction.
- **metallic** – ***non-directional, electron sharing***, which gives an electron cloud that is responsible for the good electrical conductivity
- **covalent** – ***highly directional, electron sharing***. Due to the high directionality of the bonds, the coordination number (CN) is typically low since open network structures are formed
- **secondary** bonds – are a result of attraction between induced (temporary) or permanent ***dipoles***.

Questions

How are bond energy and bond force related?

What is the type of bonding of quartz, SiO_2?

Why is ionic bonding in NaCl energetically favorable?

What is the type of bonding found in diamond?

What causes the presence of dipoles?

3 Crystal Structures

Most engineering materials are **crystalline**, which means that the atoms of the material are arranged in a **regular** and **repeating** structure. Common to all the crystalline materials are the fundamentals of crystal geometry. We will identify:

- **7 crystal systems** *and*
- **14 crystal lattices.**

3.1 Crystal Systems and Lattices

The central feature of a crystalline structure is that it is a **regular** and **repeating** arrangement of atoms, as shown for the two crystal structures of **Mg** and **Al** in Figure 3.1.

To quantify the specific repetition, one has first to decide what **structural unit** is being repeated. Any crystalline structure could be described as a pattern formed by repeating various structural units (Figure 3.2).

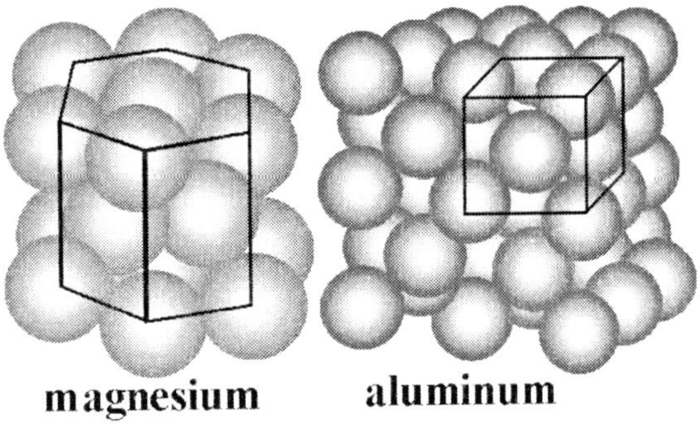

Fig. 3.1 Crystal structures of Mg (hexagonal) and Al (cubic).

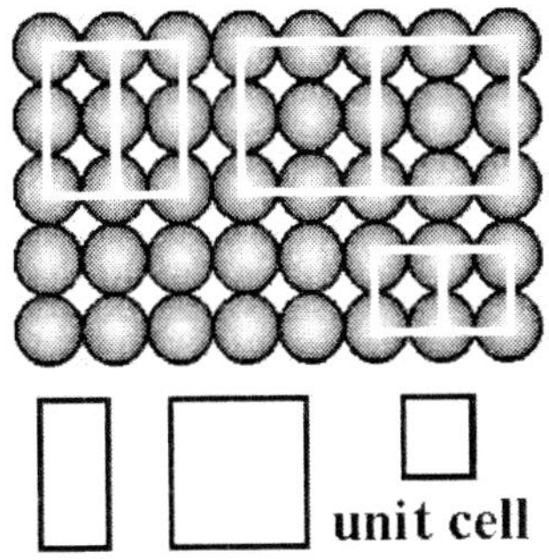

Fig. 3.2 Possible repeat units for the same atomic arrangement.

However, only the most simple repeat unit is referred to as the **unit cell** of the crystal structure. The general geometry of a unit cell is characterized by a set of lengths and angles (the so-called *metric*; see Figure 3.3).

The key feature of the unit cell is that it contains a **full description** of the structure of the material as a whole, because the crystal structure can be generated by a *repeated stacking* of adjacent unit cells face to face throughout three dimensions.

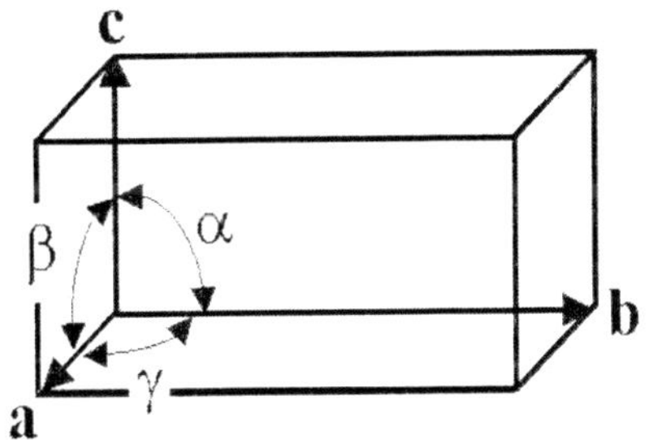

Fig. 3.3 Each unit cell is characterized by a given set of lengths (a, b, c) and angles (α, β, γ); see also Table 3.1.

The **unit cell concept** has the main advantage that all possible structures can in fact be reduced to a rather *small number* of basic unit cell geometries (Table 3.1). As a consequence, there are only **seven crystal systems** defined.

Tab. 3.1 Axial lengths and angles of the seven crystal systems.

Crystal System	Axial Lengths and Angles
cubic	$a = b = c \quad \alpha = \beta = \gamma = 90^\circ$
tetragonal	$a = b \neq c \quad \alpha = \beta = \gamma = 90^\circ$
orthorhombic	$a \neq b \neq c \quad \alpha = \beta = \gamma = 90^\circ$
rhombohedral	$a = b = c \quad \alpha = \beta = \gamma \neq 90^\circ$
hexagonal	$a = b \neq c \quad \alpha = \beta = 90^\circ$, $\gamma = 120^\circ$
monoclinic	$a \neq b \neq c \quad \alpha = \gamma = 90^\circ \neq \beta$
triclinic	$a \neq b \neq c \quad \alpha \neq \beta \neq \gamma \neq 90^\circ$

The length of unit cell edges and the angles between crystallographic axes are referred to as **lattice constants** or **lattice parameters**.

After introducing the overall geometrical arrangements of the different crystal systems, one now has to consider how atoms or ions can be **stacked together** within a given unit cell. To do this, one considers **lattice points**, which are *theoretical points* arranged in a periodic manner in three dimensions. ***Note*** that crystals are built by placing single atoms or groups of atoms (or molecules) at the lattice points of the respective crystal structure.

Again, there is a rather limited number of possibilities to define lattice points for each crystal system. ***Note*** that for some crystal systems there is more that one possibility, represented by the **fourteen Bravais**[6] **lattices** (Figure 3.4).

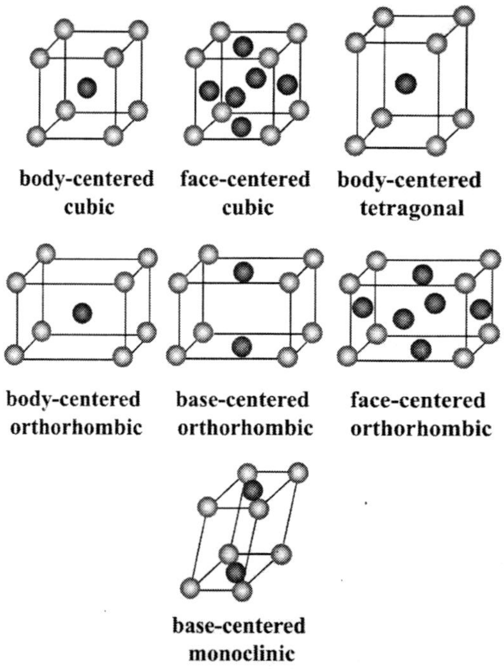

Fig. 3.4 Shown are the *additional* seven non-primitive Bravais lattices. *Note* that the primitive Bravais lattices are identical to the seven crystal systems given in Table 3.1 (here, the additional atoms are shown slightly darker).

3.2 Lattice Positions, Directions and Planes

There are a few basic rules for describing the geometry in and around the unit cell. The associated notation is *uniformly* used by crystallographers and geologists as well as material scientists and engineers.

Within a given structure one has to distinguish between **(i) atom positions**, **(ii) crystal planes**, and **(iii) crystal directions** (Figure 3.5).

[6] **Auguste Bravais** (1811-1863), French crystallographer.

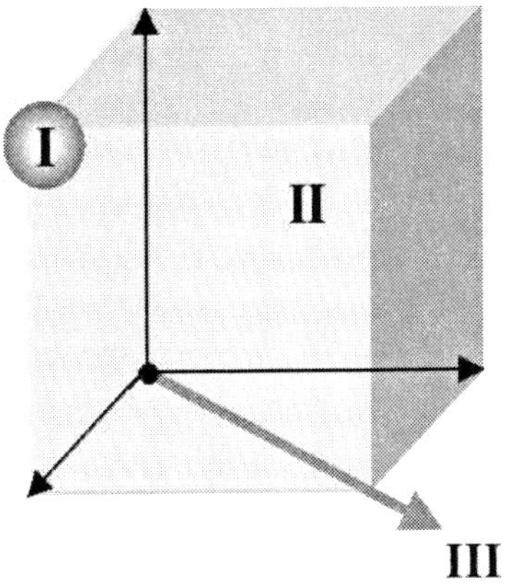

Fig. 3.1 Schematic representation of (i) atom position, (ii) crystal plane, and (iii) crystal direction.

3.2.1 Lattice Positions

Lattice positions are expressed as **fractions** or **multiples** of unit cell dimensions. For example, the **body-centered position** (see Bravais lattices) in the unit cell projects midway along each of the three unit edges (half the diagonal) and is designated the position ½ ½ ½ (Figure 3.6).

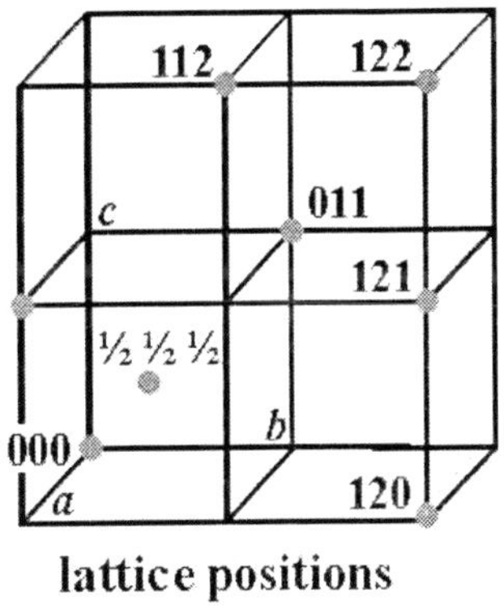

Fig. 3.6 Various lattice positions of a body-centered cubic lattice.

One important aspect of the nature of crystalline solids is that the given lattice position in a given unit cell is **equivalent** to the same position in any other unit cell of the same structure, hence, one needs only to describe one lattice position to characterize the whole crystal structure (with respect to that position).

These **equivalent positions** are connected by lattice translations, consisting of integral **multiples** of lattice constants along directions parallel to one of the three crystallographic axes (*a*, *b*, *c*) or any other direction such as [111] or [210].

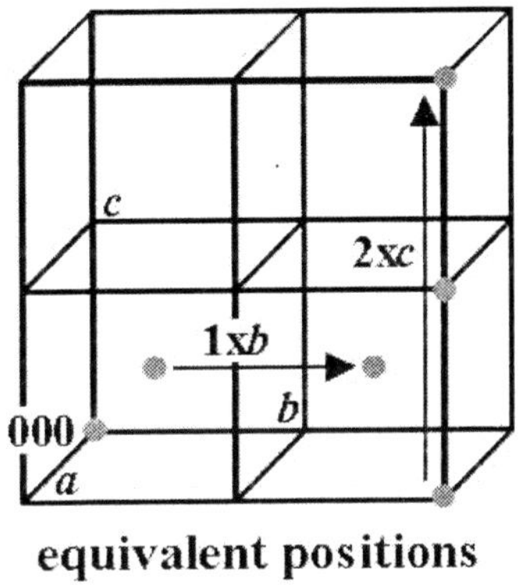

Fig. 3.7 Equivalent lattice positions in a body-centered cubic lattice.

3.2.2 Lattice Directions

Lattice directions are *always* expressed in **sets of integers**, which are obtained in a straightforward way: one identifies the smallest integer position **intercepted** by a line from the origin of the crystallographic axes to the corresponding lattice position.
To distinguish a lattice direction from a lattice position, one encloses the direction integers in **square brackets**. When a direction moves along a *negative* axis, this is indicated by a **bar** above the integer: $[\bar{1}11]$ (*bar one, one, one*), which corresponds to a line from the origin intersecting the −111 position.

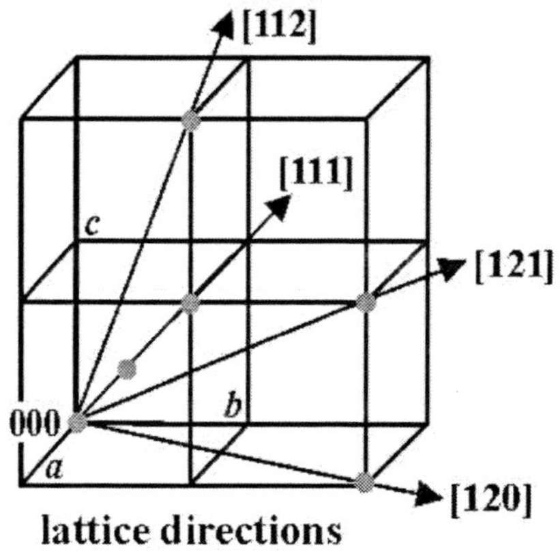

Fig. 3.8 Lattice directions in a body-centered cubic crystal.

If one compares the $[111]$ and the $[1\bar{1}\bar{1}]$ direction in the cubic system, one would find that the $[1\bar{1}\bar{1}]$ direction becomes [111] simply by choosing a different origin (choice of crystallographic axis). Such a set of directions is *structurally equivalent* and is termed a **family of directions**, designated by **angular brackets <111>**.

For example, the body diagonal in a cubic system can be described as:

<111> $=[111][1\bar{1}1][1\bar{1}1][11\bar{1}][\bar{1}\bar{1}1][1\bar{1}\bar{1}]$
$[\bar{1}1\bar{1}][\bar{1}\bar{1}\bar{1}]$; $\Sigma=8$

Sometimes, in particular when dealing with mechanical properties, it can be useful to know the angle between two given directions.

In the **cubic** system, the angle between directions can be determined by the dot-product of two vectors. Taking two directions **[uvw]** and **[u'v'w']** as vectors

$$\vec{D}=(u\cdot\vec{a})+(v\cdot\vec{b})+(w\cdot\vec{c}) \quad and$$
$$\vec{D}'=(u'\cdot\vec{a})+(v'\cdot\vec{b})+(w'\cdot\vec{c}),$$

the angle δ becomes:

$$\vec{D}\cdot\vec{D}'=|D|\cdot|D'|\cdot\cos\delta$$
$$\cos\delta=\frac{\vec{D}\cdot\vec{D}'}{|D|\cdot|D'|} \qquad (3.1)$$
$$=\frac{u\cdot u'+v\cdot v'+w\cdot w'}{\sqrt{u^2+v^2+w^2}\cdot\sqrt{(u')^2+(v')^2+(w')^2}}$$

Note that the above equation 3.1 *only applies* for the **cubic** system. For other crystallographic systems, the corresponding equations are more complex.

3.2.3 Lattice Planes

As for directions, lattice planes are expressed as a **set of integers**, known as **Miller indices**[7].

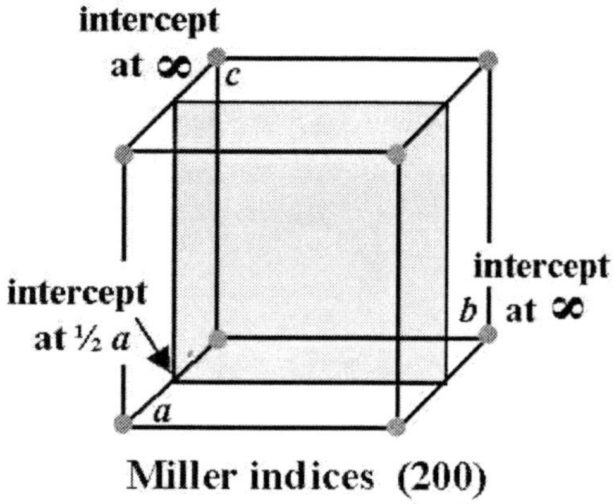

Fig. 3.9 Miller indices of a lattice plane that only intercepts at ½ *a*.

These integers represent the **inverse** of the three axial intercepts. As an example, let us consider the **(210) plane**. It intercepts the *a*-axis at **½ *a***, the *b*-axis at ***b*** and runs parallel

[7] **William Hallowes Miller** (1801-1880), British crystallographer.

to the c-axis (in effect, intercepting it at **infinity, ∞**).

The inverse of the intercepts are $\frac{1}{1/2}, \frac{1}{1}, \frac{1}{\infty}$,

which gives 2, 1, and 0 leading to the **(210) notation.** The general notation of Miller indices is **(hkl)** and can be used for any of the other crystal systems. *Note* that in the cubic system, the [hkl]-direction is *always* normal to the (hkl)-plane.

Because the **hexagonal** system can be conveniently presented by four axes, a 4-digit set of **Miller-Bravais indices, (hkil)**, is defined, with $i = -(h+k)$.

Miller indices = (210)
Miller Bravais indices = $(21\bar{3}0)$.

Note that converting directions from the Miller notation into Miller-Bravais indices (hexagonal system) is not as straightforward as for planes.

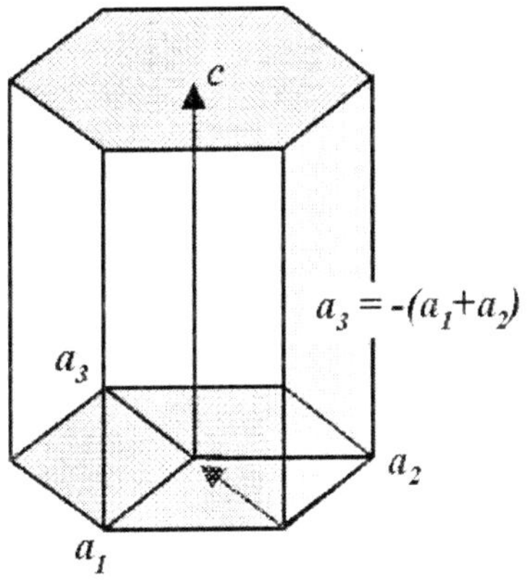

Fig. 3.10 The hexagonal system.

As for structurally equivalent directions, we can group *structurally equivalent planes* as a **family of planes** by enclosure in **braces {100}**.

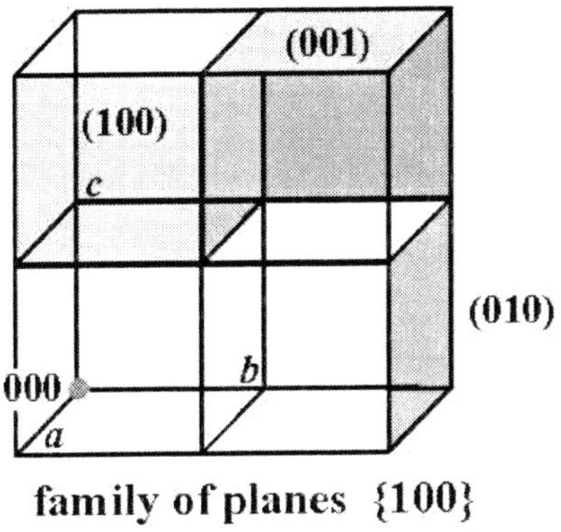

Fig. 3.11 Family of {100} planes.

In the cubic system, the {100} family is:
{100} = $(100)(010)(001)(\bar{1}00)(0\bar{1}0)(00\bar{1})$

Summary of unit cell elements

- **7 crystal systems**
- **14 Bravais lattices**

One distinguishes between:

- lattice **points**
- lattice **directions**
- lattice **planes**

- Lattice **Points**

Set of multiples or fractions of lattice vectors; 1 0 ½.

- Lattice **Directions**

Set of integers where the direction *intersects* with a lattice vector. Brackets [111]; family of directions – angular brackets <100>.

- Lattice **Planes**

Set of integers where the lattice plane *intersects* with lattice vectors. The inverse of the intersections gives the Miller indices (hkl) written in parenthesis (110); a family of planes is given in braces {110}. For the hexagonal system, however, Miller-Bravais indices (hkil), with $i = -(h+k)$ are commonly used.

3.3 Metal Structures

In the following sections, the main crystal structures associated with important engineering materials are introduced. For the first important group, elemental metals, the number of common crystal structures is fairly small, since most metals at room temperature are found in one of three crystal structures:

- **body-centered cubic (bcc)**
- **face-centered cubic (fcc)**
- **hexagonal close packed (hcp)**

3.3.1 Body-Centered Cubic

The body-centered cubic structure can be described as a **bcc Bravais lattice** with **one atom** centered at each lattice point. There is one atom at ½ ½ ½ and one atom at each unit cell **corner**. Each corner atom is **shared** with 8 adjacent unit cells. Thus there are **2 atoms** in the bcc unit cell: 1 + 8x1/8 = 2.

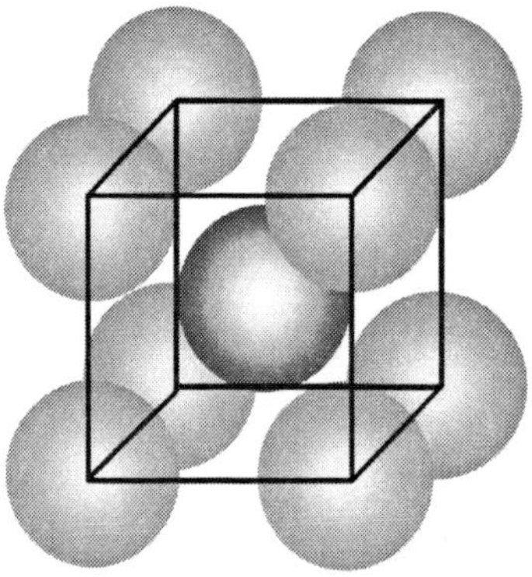

Fig. 3.12 Body centered cubic crystal structure.

The **atomic packing factor** (APF) for the **bcc** structure is **0.68**, which represents the fraction of the **unit cell volume** occupied by the two metal atoms. Note that the packing density of the simple (primitive) cubic system is only 0.52.

Typical **bcc metals** are: **α-Fe, K, Ti, V, Cr, Mo, Nb, Ta** and **W**.

Note that most alloys with one of the above metals as the predominant constituent will also reveal the bcc structure.

3.3.2 Face-Centered Cubic

The face-centered cubic structure can be described as an **fcc Bravais lattice** with **one atom** per lattice point. There is one atom on each **face** (shared by 2 unit cells) and one atom at each **corner** (shared by 8 neighboring unit cells). The total number of atoms equals **4** (6x1/2 + 8x1/8 = 4), which occupy **0.74** (APF) of the unit cell volume. In fact, the APF of **0.74** is the **highest value** for filling space by stacking equal-sized hard spheres. Therefore, the fcc structure is sometimes referred to as **cubic close packed**.

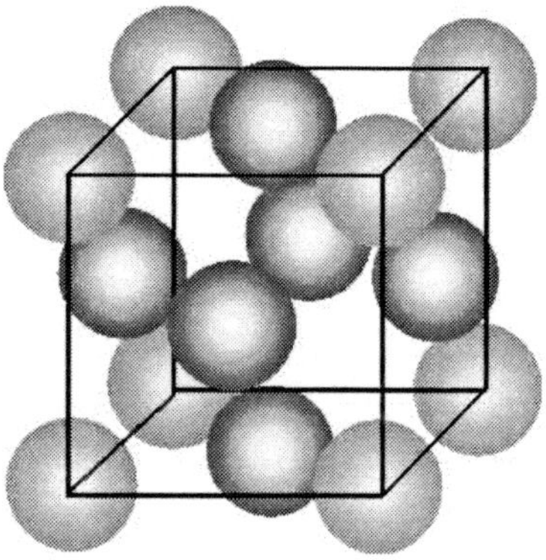

Fig. 3.13 Face centered cubic crystal structure.

Typical **fcc metals** are: **Al**, **Ni**, **Cu**, **Pb**, **Ag**, **Au**, **Pt**, and **γ-Fe** (912-1394 °C).

3.3.3 Hexagonal Close Packed

The hexagonal close packed structure is the first crystal structure we encounter that is **more complex**. There is **one atom centered within** the unit cell and **various atoms** at the unit cell corners that are shared by 6 or 12 adjacent unit cells (4x1/6 + 4x1/12), giving a total of **2 atoms** per unit cell: 1 + 4x1/6 + 4x1/12 = 2.

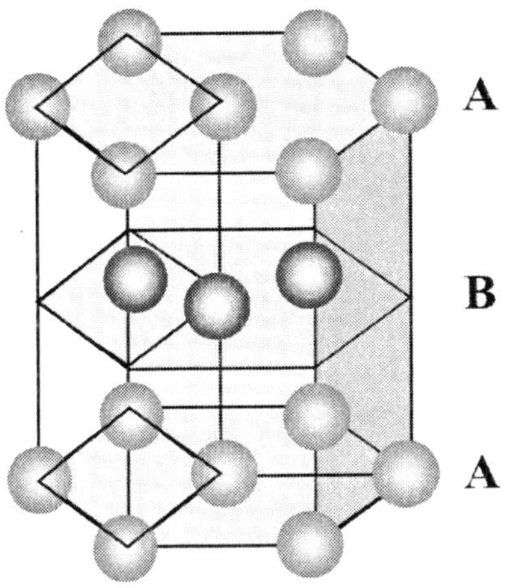

Fig. 3.14 Hexagonal close packed structure.

Typical **hcp metals** are: **Be, Mg, α-Ti, Zn, Co, Cd,** and **Zr**.

The hcp structure (Figure 3.14) also has an APF of **0.74**. The fact that this structure shows the same AFP as the **fcc** structure raises the question: **what is the basic difference or similarity between fcc and hcp?**

It is simply the difference in **stacking sequence**. Both structures are regular packing sequences of close packed planes. In **fcc**, the **(111) planes** are close packed. These planes are **identical** to the **(0002) planes** of the **hcp** structure.

In the **fcc** arrangement, every **4th (111)** layer lies precisely above the first (Figure 3.15). As a consequence, the stacking sequence is **ABC**ABCABC.

In the **hcp** arrangement, every **3rd (0002)** layer lies precisely above the first, which results in a stacking sequence of **AB**ABAB (compare Figures 3.14 and 3.15).

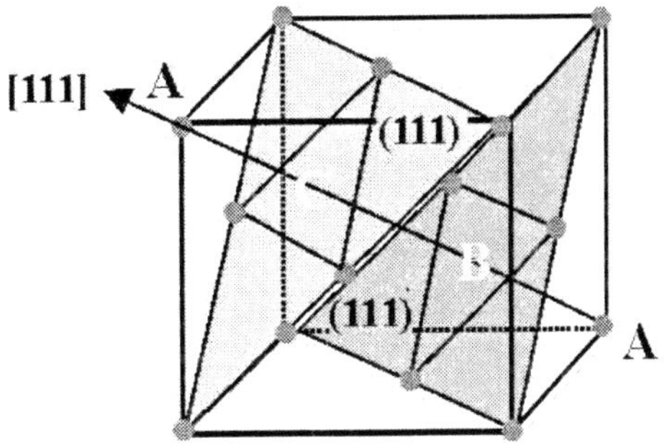

Fig. 3.15 Stacking sequence **ABC**ABC of the fcc structure with an APF of 0.74. *Note* that the hcp structure shown in Figure 3.14 has the same APF but a different stacking sequence (**AB**ABAB).

Note that there are exceptions to the rule that all metals crystallize in one of the *three metal structures* just discussed (bcc, fcc, hcp). One example is **α-Uranium**, which has a **body-centered orthorhombic** Bravais lattice with **2 atoms** associated with each lattice point.

3.4 Ceramic Structures

The wide variety of compositions of ceramic compounds is also reflected in their numerous crystal structures. ***Note*** that many of the ceramic structures also describe the structure of intermetallic phases.

NaCl

Fig. 3.17 The NaCl structure.

3.4.1 MX-Compounds

MX-compounds are ceramic structures where **M** is a **metal** and **X** is a **non-metallic** element. A well known representative of this group is **CsCl**.

At first glance this structure looks just like a **bcc Bravais** lattice, however, here we have ***two types if ions***. Therefore, the CsCl structure represents a **simple cubic** Bravais lattice, where two cubic *sublattices* are *interpenetrating*. The CsCl structure has **2 atoms** in the unit cell: 1 + 8x1/8 = 2.

CsCl

Fig. 3.16 The CsCl structure.

Rocksalt, **NaCl**, is also a **MX-compound** and its structure is shared by many important ceramics like **MgO**, **CaO**, **FeO**, and **NiO**. The structure can be viewed by the intertwining of two fcc lattices. There are **8 atoms** in the unit cell: 8x1/8 + 12x1/4 + 6x1/2 + 1 = 8.

3.4.2 MX_2-Compounds

The **MX_2 structure** also includes a large number of important ceramics such as **CaF_2**, **UO_2**, **ThO_2**, and **TeO_2**.

The **CaF_2 – Fluorite – structure** is built on an **fcc** Bravais lattice with **3 atoms** associated with each lattice point (fcc = 4(x3)=12). There are **12 ions** per unit cell (**4 Ca^{2+}** and **8 F^-**).

CaF_2

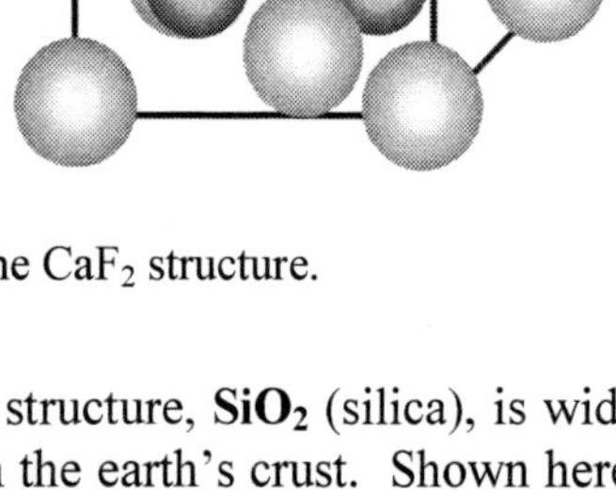

Fig. 3.18 The CaF_2 structure.

The quartz structure, **SiO_2** (silica), is widely available in the earth's crust. Shown here is the cubic **fcc** structure of the high-temperature stable **polytype cristobalite** (1470-1723 °C; see also Table 3.2) with **6 ions** (**2 Si^{4+}** and **4 O^{2-}**) associated with each lattice point. Hence, there are **24 ions** per unit cell: 1/8(8x6) + 1/2(6x6) = 24.

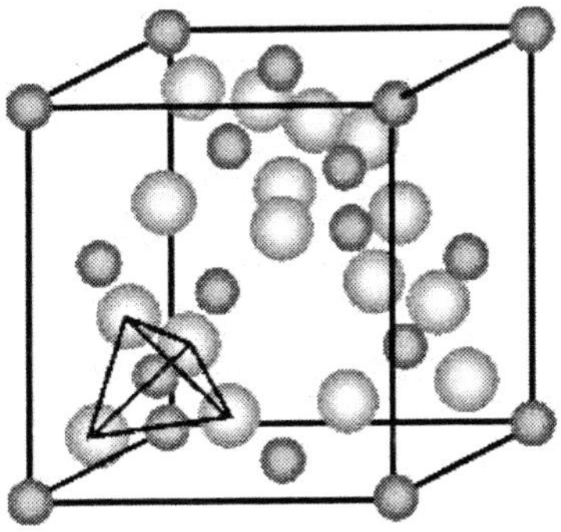

Fig. 3.19 The SiO_2 structure (cristobalite).

Note that the overall silica structure is a **continuously connected network** of SiO_4^{4-} **tetrahedra** with **shared O^{2-} ions** between them, which results in the overall formula **SiO_2**. Silica reveals mixed bonding (ionic plus covalent = 50:50).

Tab. 3.2 Various polytypes of silica with their corresponding crystal system and point group.

Mineral	Crystal System	Point Group
α-Quartz	trigonal	3 2
α-Tridymite	orthorh. *or* monoclinic	2/m 2/m 2/m *or* 2/m
α-Cristobalite	tetragonal	4 2 2
β-Quartz	hexagonal	6 2 2
β-Tridymite	hexagonal	6/m 2/m 2/m
β-Cristobalite	isometric	4/m –3 2/m
Coesite	monolcinic	2/m
Stichovite	tetragonal	4/m 2/m 2/m
Lechatelierite	amorphous	*none*
Keatite	tetragonal	4 2 2

3.4.3 M_2X_3-Compounds

The M_2X_3 structure includes the important mineral **corundum**, **Al_2O_3**, which is also known by ceramists as **alumina**. It is a **rhombohedral Bravais** lattice, but closely approximates a **hexagonal** lattice.

There are **30 ions** in the unit cell (**12 Al^{3+}** and **18 O^{2-}**). One can visualize this rather complex structure being similar to the **hcp** structure, because the **O^{2-}** ions form **close packed planes** (sheets) that are stacked upon each other (Figure 3.20). However, only **2/3** of the **interstices** (void space) between these sheets are filled with the **Al^{3+}** ions.

Typical representatives that crystallize in the **corundum structure** are: **Al_2O_3**, **Cr_2O_3**, and **α-Fe_2O_3** (hematite=rust).

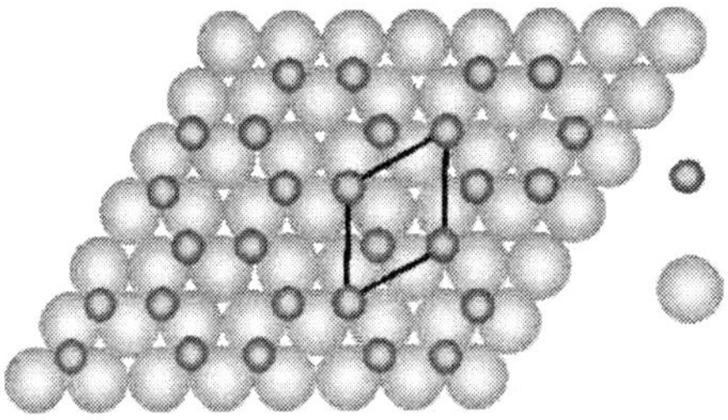

Fig. 3.20 Al_2O_3 crystal structure (corundum).

3.4.4 ABX_3-Compounds

The ABX_3 structure includes the important family of **electronic ceramics** which all show the **perovskite**, **$CaTiO_3$**, structure.

At first glance, this structure appears to be a combination of **simple cubic** with **bcc** and **fcc**. However, the **Ca^{2+}** ions occupy the **corners** of the unit cell, **Ti^{4+}** is placed at the **body center** and the **O^{2-}** ions occupy the **face-centered** positions. As a consequence, the perovskite structure is an example of a **simple cubic** Bravais lattice with three different types of ions within the unit cell. There are **5 ions** per unit cell (**1Ca^{2+}**, **1Ti^{4+}**, **3O^{2-}**); 1+1+6x1/2 = 5.

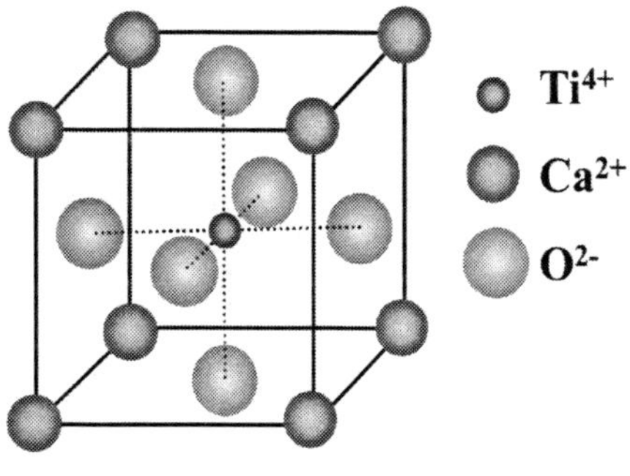

Fig. 3.21 $CaTiO_3$ crystal structure (perovskite).

Note that **$BaTiO_3$** is an important electronic ceramic with interesting **ferroelectric** and **piezoelectric** properties, which reveals the perovskite structure.

3.4.5 AB_2X_4-Compounds

The AB_2X_4 structure includes an important family of **magnetic ceramics**, which show the **spinel, $MgAl_2O_4$**, structure.

This rather complex structure is built on an **fcc** Bravais lattice with **14 ions** associated with each lattice point (**2Mg^{2+}**, **4Al^{3+}**, **8O^{2-}**). There are **56 ions** within the unit cell (**8Mg^{2+}**, **16Al^{3+}**, **32O^{2-}**). In the regular spinel structure (Figure 3.22), the **M^{2+}** ion sits in **tetrahedral** coordination, while the **M^{3+}** ion is **octahedrally** coordinated: $M^{2+}M_2^{3+}O_4$.

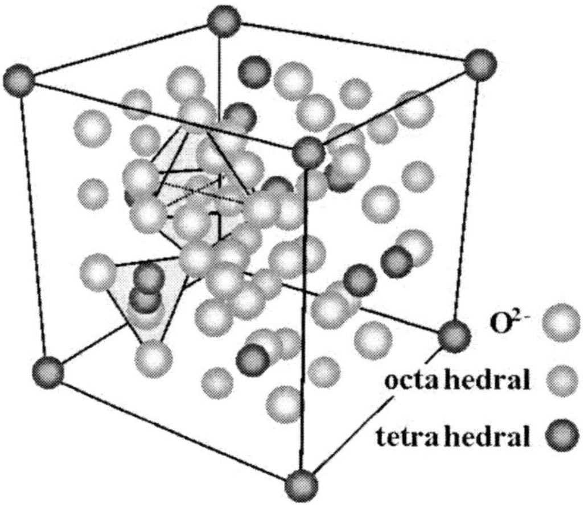

Fig. 3.22 $MgAl_2O_4$ crystal structure (spinel).

Typical compounds that reveal this structure are **$NiAl_2O_4$**, **$ZnAl_2O_4$**, **$FeMgFeO_4$**, **$FeNiFeO_4$** and **$ZnFe_2O_4$** (note that one can simply replace the ion species).

Apart from the regular spinel structure, there is a specific type of the spinel structure, the **inverse spinel.** Here, the **octahedral** sites are occupied by all **M^{2+}** ions and ½ of the **M^{3+}** ions and the **remaining M^{3+}** ions are filling the **tetrahedral** sites, which leads to the formula: $M^{3+}(M^{3+}, M^{2+})_2O_4$.

Ceramic magnets are based on the **inverse spinel** structure: **$FeMgFeO_4$**, **$FeFe_2O_4$** (**Fe_3O_4 = magnetite**), **$FeNiFeO_4$ = ferrites.**

Note that there are many more complex structures found in nature as for example the structure of different clay minerals. One example is **kaolinite**, **$Al_4Si_4O_{10}(OH)_8$**, which is a **sheet silicate** with a **triclinic** Bravais lattice and 2 kaolinite "molecules" per unit cell.

3.5 Semiconductor Structures

The elemental semiconductors **Si** and **Ge** as well as **gray Sn** all share the cubic **diamond structure**, which is built on an **fcc** Bravais lattice with 2 atoms associated with each lattice point and **8 atoms** per unit cell (Figure 3.23). The key feature of this structure is that it establishes a **tetrahedral bonding configuration** (sp^3-hybrid orbitals; see also atomic bonding section).

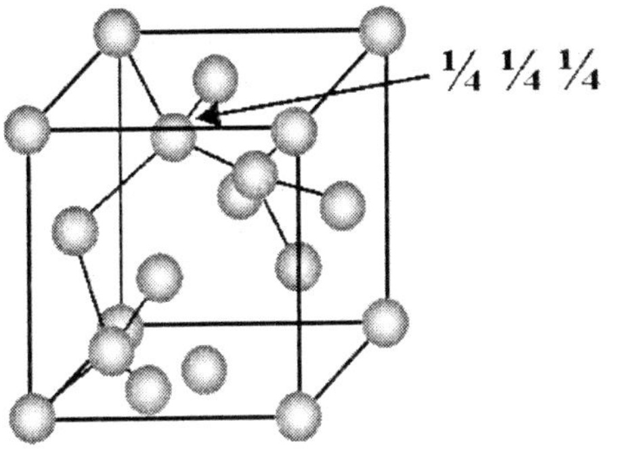

Fig. 3.23 The structure of Si is shown which is based on the diamond structure.

Many of the simple MX semiconductor structures such as **II-VI** and **III-V** compounds (average valence = 4^+) crystallize in the zinc blende, **ZnS**, structure, which is closely related to the diamond structure, but with $\mathbf{Zn^{2+}}$ and S^{2-} *alternating* in the atom positions. As in diamond, there are **8 ions** ($4Zn^{2+}$ and $4S^{2-}$) per unit cell (Figure 3.24).

Representatives of the **zinc blende structure** are: **ZnSe**, **CdS**, **HgTe** (II-VI) and **GaAs**, **AlP**, **InSb** (III-V).

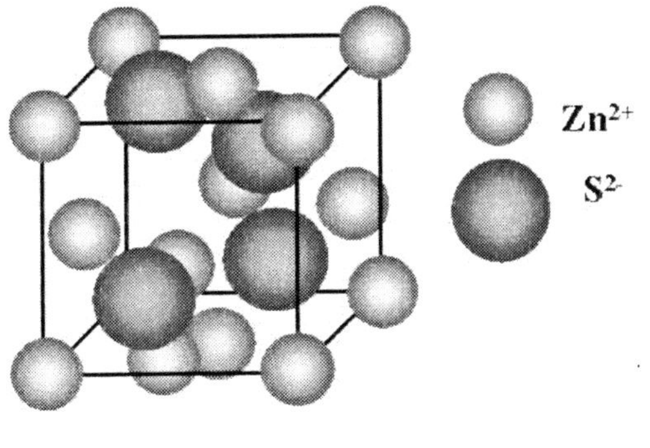

Fig. 3.24 The zinc blende, ZnS, structure with alternating Zn- and S-ions.

3.6 X-Ray Diffraction

In sections 3.1-3.5 we introduced how crystalline structures can be described and what some specific crystal structures look like. We will summarize the chapter on crystal structures by learning how one can determine in which structure a given compound crystallized. The technique that is commonly used for crystal structure and phase identification is **X-ray diffraction.**
This technique can generally be used to determine the crystal structure of a newly synthesized compound or it can be employed to characterize the structure of an unknown sample (but with already known structure). X-ray diffraction (XRD) patterns can be obtained in a number of different ways. Two well-known techniques are:

- **Laue camera**[8]
- **Powder diffractometer**.

Diffraction is a consequence of the scattering of radiation by a regular array of scattering centers like a grating or a row of atoms. Hereby it is assumed that the **distance** of the scattering centers is on the order of the **wavelength** of radiation.

---

[8] **Max von Laue** (1879-1960), German physicist.

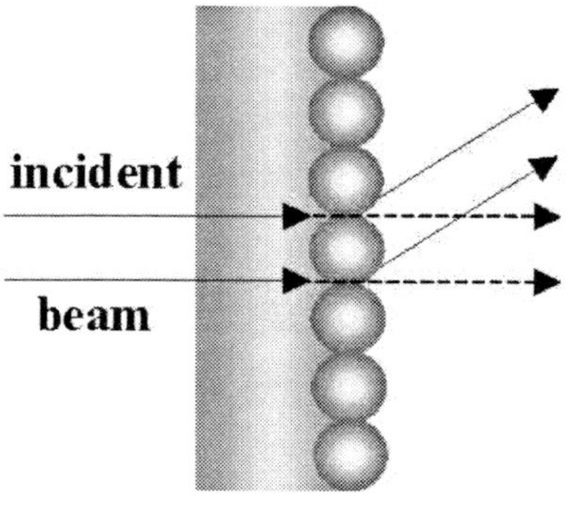

Fig. 3.25 Diffraction grating; either parallel scratches on a glass surface or a row of atoms can act as individual scattering centers.

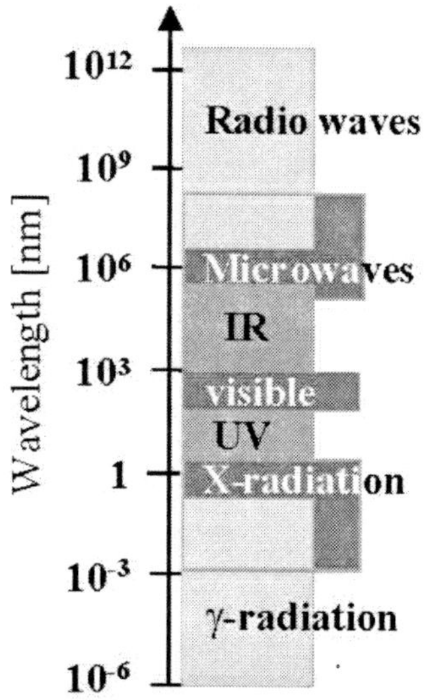

Fig. 3.26 Wavelengths of the electromagnetic radiation spectrum.

Parallel scratch lines placed in a regular manner about 1 μm apart cause diffraction of visible light. We can think of a single crystal as a special grating with a separation distance of the **scattering centers** (atoms or ions; Figure 3.25) of only a few tenths of a nm (lattice parameter: a_0 of bcc-Fe = 0.2868 nm).

Note that since one requirement for diffraction is that the wavelength of the radiation be on the order of the spacing between the scattering centers, it is obvious that when using a shorter wavelength (as compared to visible light) smaller details (atomic distances) can be resolved.

The incoming radiation is scattered with a strong intensity in just a few **specific directions**. This process is due to **destructive** and **constructive interference** of the scattered electromagnetic waves.

Looking at the **electromagnetic spectrum**, it becomes obvious that when using a **smaller wavelength** of radiation (X-rays, electrons, Figure 3.26), it is possible to resolve smaller details (atomic distances) of matter. Therefore, X-ray diffraction has become a very powerful tool for materials characterization.

The specific mechanism of X-ray scattering is the interaction of a **photon** of electromagnetic radiation with the **electrons** of the scattering atom. The crystal acts as a **3-dimensional diffraction grating** with a repeated stacking of equivalent lattice planes (close packed planes).

One very important **requirement** has to be fulfilled in order for **diffraction** to occur (constructive interference): the electro-magnetic waves diffracted by adjacent atomic planes have to be **in phase**; otherwise **destructive interference** is the result and there will be no scattered radiation to be recorded (see also XRD pattern in Figure 3.28 and extinction rules in Table 3.3).

At the precise geometry of **constructive interference** (in-phase scattering), the **difference** in **path length** between the adjacent X-ray beams is some integral number, n, of the radiation wavelength, λ. The relationship that describes this dependence is called the **Bragg equation**[9]:

$$n\lambda = 2d \cdot \sin\Theta \qquad (3.2)$$

where d is the **lattice spacing** between adjacent lattice planes and Θ is the **scattering angle**. The scattering angle is usually referred to as the **Bragg angle** and the angle 2Θ is called the **diffraction angle**, which can easily be determined experimentally (X-ray diffraction; Figure 3.27).

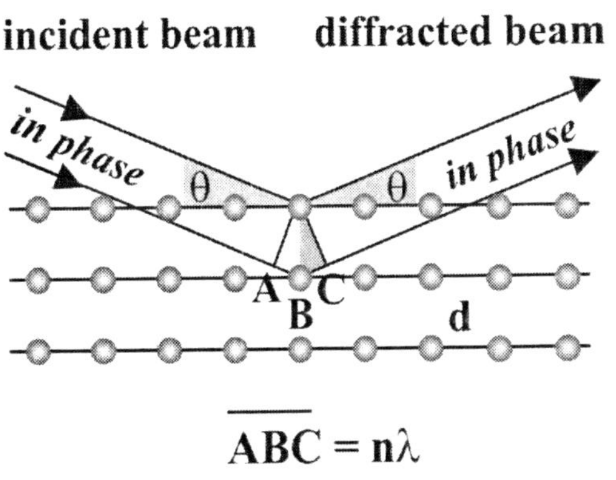

Fig. 3.27 Geometric representation of Bragg's law: 2dsinΘ=nλ.

The distance of **interplanar spacing** within a given crystal structure is directly proportional to the **Miller indices** (hkl) of a specific plane (distance between adjacent planes).

For the **cubic** system this correlation is rather simple. The spacing between adjacent (hkl) planes can be calculated by equation 3.3:

$$d_{hkl} = \frac{a_0}{\sqrt{h^2 + k^2 + l^2}} \qquad (3.3)$$

where a_0 is the lattice parameter (edge length of the unit cell).

When combining equations 3.2 and 3.3 it follows:

$$\sin^2\Theta = \frac{\lambda^2}{4a_0^{\,2}}\left(h^2 + k^2 + l^2\right) \qquad (3.4)$$

Note that the above given **Bragg equation** is **only valid** for simple or so-called **primitive Bavais lattices**; that is crystal lattices that only reveal atoms at the corners of the unit cell.

Lattices that do contain **additional atoms** or ions like bcc or fcc (at the face of the unit cell or in the center of the body), or even located along a unit cell edge (1 ½ 0), are the so-called **non-primitive** unit cells. These additional scattering centers can cause **destructive, out-of-phase interference**, which results in a lowered or even **extinct intensity** of the reflected radiation. Some of the **extinction rules** that describe this phenomenon are given in Table 3.3.

---

[9] **William Henry Bragg** (1862-1942) *and* **William Lawrence Bragg** (1890-1971), English physicists (father and son) who described the interaction between electro-magnetic wavelength and lattice spacing in 1912.

Tab. 3.3 Extinction rules for bcc, fcc, and hcp lattices.

Crystal Structure	Extinction	Diffraction
bcc	**h+k+l = odd**	**h+k+l = even**
fcc	**h,k,l = mixed (even and odd)**	**h,k,l unmixed (all even or all odd)**
hcp	**(h+2k) = 3n, l=odd#**	**all other combinations**

for example (221); n = integer

From the X-ray diffraction spectrum shown in Figure 3.28 one can see that not all possible reflections were monitored. Al reveals the fcc structure and, therefore, all reflections of mixed character are not seen, as for example (210) or (100).

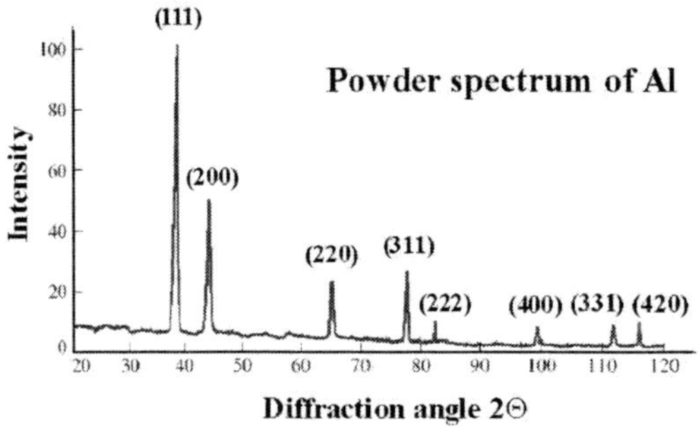

Fig. 3.28 X-ray diffraction spectrum of an Al-powder sample.

Example 3.1

Calculate the d-spacing (unit cell length, a_0) of fcc Al metal using Bragg's equation. Given is the reflection (400) at $99^0 = 2\,\Theta$; use the wavelength of radiation $\lambda = 0.1542$ nm.

$$n\lambda = 2d \cdot \sin\Theta$$

$$4 \cdot 0.1542 nm = 2d \cdot \sin(99/2)$$

$$d = 0.405\, nm$$

4 Crystal Defects

Before, we have exclusively dealt with perfectly repetitive crystalline solids. However, nothing is perfect. No material exists that does not exhibit any **imperfection**, which might arise during processing of the material or during grain growth or secondary phase precipitation (annealing).

We will distinguish the following crystal **defects/imperfections**:

4.1 chemical imperfection
- random vs. ordered
- non-stoichiometric compounds

4.2 point defects *(zero-dimensional)*
- vacancy
- interstitialcy
- Schottky defect
- Frenkel defect

4.3 linear defects *(one dimensional)*
- dislocations
  - ❑ edge dislocation
  - ❑ screw dislocation
  - ❑ dislocations of mixed character

4.4 planar defects *(two dimensional)*
- grain boundaries
- twin boundaries
- tilt/twist boundaries
- amorphous grain-boundary films

4.1 Chemical Imperfections

It is nearly impossible to prevent some **contamination** during materials processing. Even high-purity semiconductors have a measurable amount of impurity atoms/ions.

When a different element is **dissolved** in a host matrix, a **solid solution** is formed (Figure 4.1). If the newly incorporated element occupies an actual **lattice site**, a **substitutional** solid solution (ss) is given.

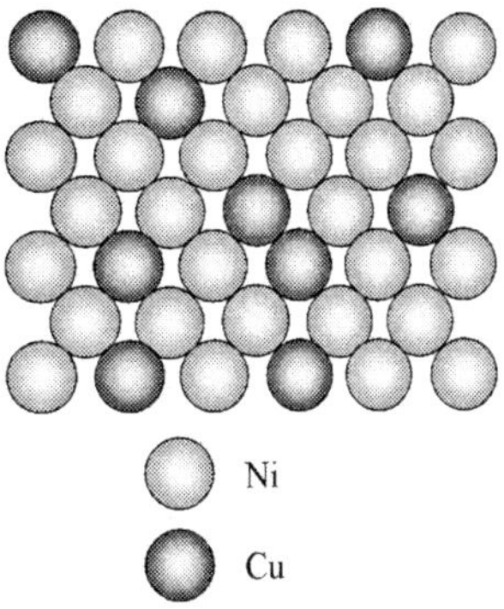

Fig. 4.1 Schematic showing the substitutional solid solution of Cu in Ni (random distribution of Cu).

This type of substitution occurs when the two atom/ion species do not differ greatly in size or crystal structure.

In metallic solid solutions, **complete miscibility** can only be expected when the two metals are very similar and, therefore, the **Hume-Rothery**[10] rules apply:

- atomic **radii difference** less than 15 %
- same **crystal structure** (bcc, fcc, or hcp)
- similar **electronegativity** (ability to attract electrons)
- same **valence**

[10] **William Hume-Rothery** (1899-1968), British metallurgist.

We have just seen the random solid solution, as in the case of **Cu/Ni-ss**. Some systems, however, form **ordered** solid solutions. One well-known example is the compound **$AuCu_3$** (Figure 4.2).

Above 390 °C this compound shows a **random** distribution of Au and Cu among fcc sites, while **below 390 °C** the Cu atoms preferentially occupy the **face-centered positions** (based on a simple cubic lattice).

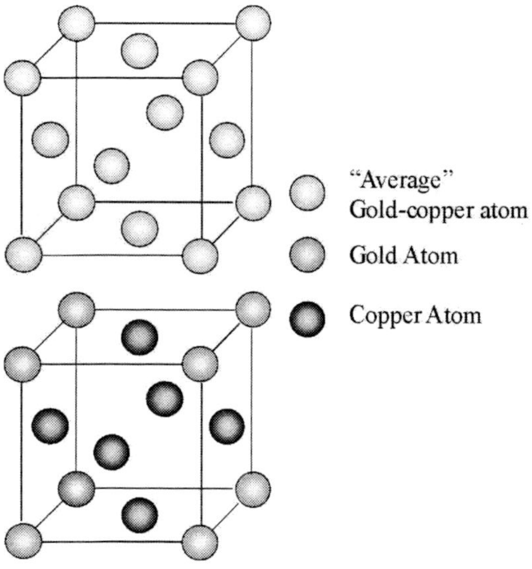

Fig. 4.2 Random versus ordered solid solution of $AuCu_3$. *Note* that calculating the number of atoms in the unit cell give the correct composition: 8x1/8=1 (Au) + 6x1/2=3 (Cu) → $AuCu_3$.

When the **atom size differs substantially**, a *substitutional solid solution* may be energetically *unfavorable*. In this case, it is more favorable for the smaller atom to go into interstitial sites, hence, creating an **interstitial solid solution**.

A typical example of this *interstitial solid solution* is the dissolution of **carbon** in bcc **α-Fe** (Figure 4.3). Although this configuration is more stable than the substitutional solid solution, the interstitial solid solution produces considerable **strain**, which greatly limits the respective **solubility** of carbon within the iron lattice to less than 0.1 %.

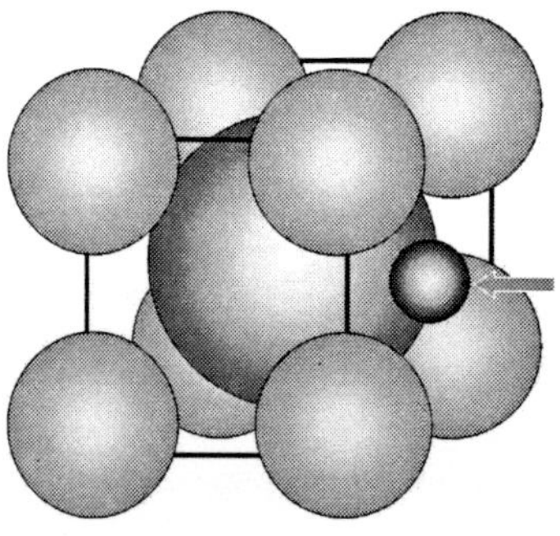

Fig. 4.3 Interstitial solid solution of carbon in α-Fe (indicated by arrow).

The *principles* of substitutional solid solution (in metals like Ni/Cu) also apply for **compounds**, as shown in Figure 4.4 for the example of **MgO/NiO** solid solution. ***Note*** that the **oxygen sublattice** is not affected by this substitution.

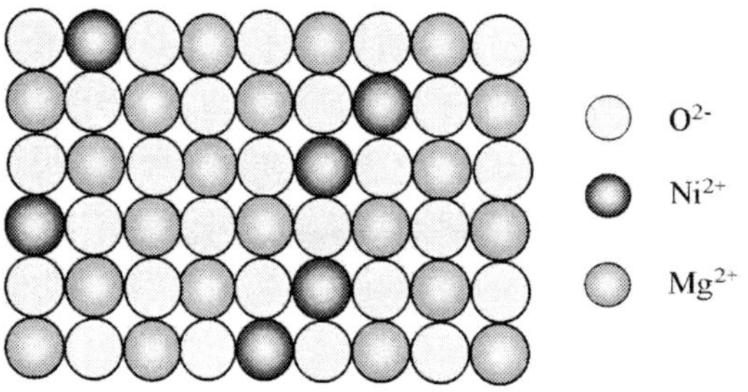

Fig. 4.4 Substitutional solid solution in compounds. Here the solid solution of NiO in MgO is shown (the oxygen sublattice is not affected by this process).

The **charge** of the dissolved ion (compound) affects the nature of the substitution.

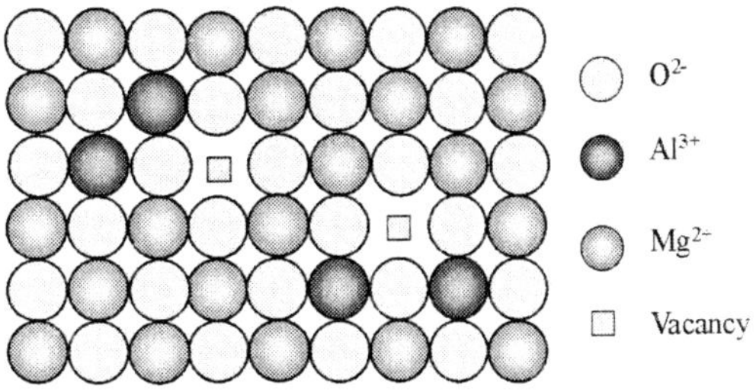

Fig. 4.5 Substitutional solid solution of Al_2O_3 in MgO. Note that in order to compensate for the charge difference, vacancies have to be created.

For example, one could not indiscriminately substitute all Mg^{2+} ions with Al^{3+} (Al_2O_3 solution). The higher valence of Al^{3+} would give a **net positive charge** to the crystal, which is **highly unstable**.

Charge neutrality has to be maintained, which is achieved by replacing only **2 Al^{3+}** for every **3 Mg^{2+}** sites (sum of charge = 6+). This leaves one **Mg^{2+}** site **vacant** for each two Al^{3+} substitutions (Figure 4.5).

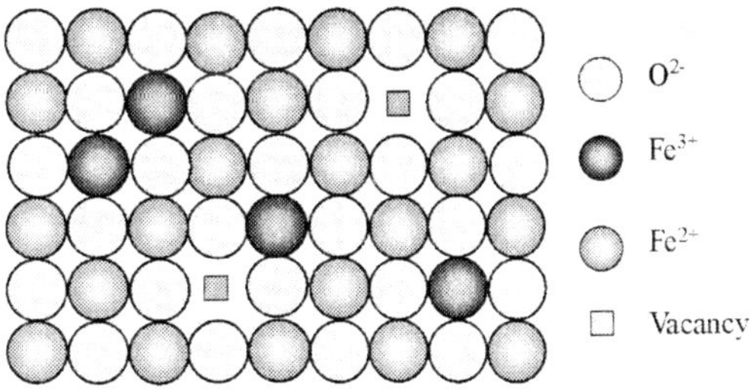

Fig. 4.6 Substitutional solid solution of Fe_2O_3 in FeO (wuestite).

Such compounds are called **non-stoichiometric** compounds and **$Fe_{1-x}O$** (wuestite) with **x=0.05** is a well known representative of this group (Figure 4.6).

An *ideally stoichiometric* **FeO** would be identical to the **MgO structure** (with only Fe^{2+}) with the NaCl-type crystal structure. However, **Fe** is a **multivalent** element in nature and there are always some Fe^{2+} oxidized to **Fe^{3+}**. As a consequence, Fe^{3+} acts similarly to Al^{3+} (Al_2O_3) in MgO. One Fe^{2+} site **vacancy** is required to **compensate** for the presence of every **2 Fe^{3+}** ions in order to generate *charge neutrality*.

4.2 Point Defects

Structural defects also exist in real materials *independent* of chemical impurities. Such defects are associated with the **crystalline point lattice** and are therefore called *point defects*. Two common types of point defects are (Figure 4.7):

- **vacancies**
- **interstitials** *or* **interstitialcy**.

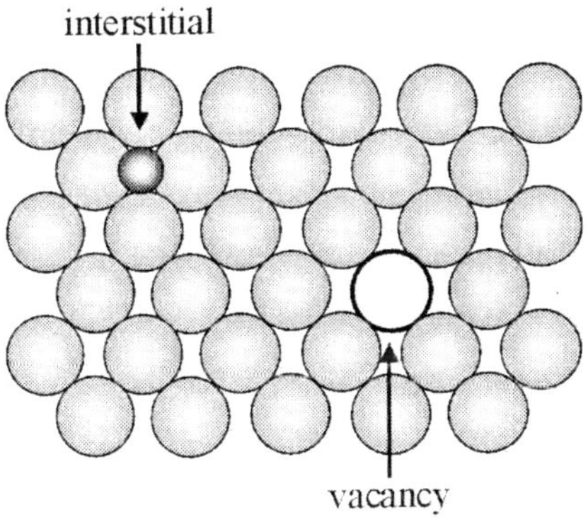

Fig. 4.7 Vacancy and interstitial point defects of a close packed lattice.

The **vacancy** is simply an **unoccupied** lattice site, while the **interstitial** is an atom occupying an **interstitial lattice site** in the perfect crystal structure (a site that is normally not occupied).

There are also **two additional** types of point defects that can bee seen as combined defects (since they represent "**pairs**"):

- **Schottky**[11] defect
- **Frenkel**[12] defect

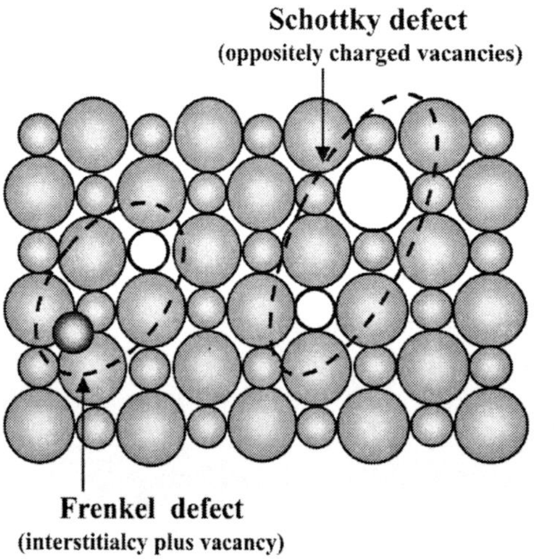

Fig. 4.8 Frenkel and Schottky defect. *Note* that both defect types are combined (paired) defects.

A **Schottky** defect is a **pair** of **oppositely charged vacancies**, while a **Frenkel** defect is a **pair** of one **vacancy** and one **interstitialcy** (Figure 4.8).

Many of the crystal structures we discussed thus far are *too densely packed* to allow Frenkel defect formation. However, the **CaF_2** structure (Figure 3.18) is relatively wide open and can accommodate vacancies and cation interstitials without *excessive lattice strain.*

Note that many defect structures can become even more complex by **local charging** owing to **electron trapping** or **electron-hole trapping**, which strongly affects the *optical properties* of the material.

4.2.1 Process Rate & Temperature

Many processes in materials science share a common characteristic: the **process rate** *rises exponentially* with temperature, as for example:

- **diffusivity** of elements in metal alloys
- **creep deformation** rate in structural materials
- **electrical conductivity** in extrinsic *n*-type semiconductors.

The general equation that describes all these various processes is:

$$process\ rate = C \cdot e^{-Q/RT} \qquad (4.1)$$

where C is a pre-exponential constant (which is *independent* of temperature), Q is the activation energy of the specific process, R is the universal gas constant, and T is the absolute temperature.

Equation 4.1 is generally referred to as the **Arrhenius equation**[13]. Taking the logarithm gives:

$$\ln(process\ rate) = \ln C - \frac{Q}{RT} \qquad (4.2)$$

A plot of the *ln* rate versus the reciprocal absolute temperature (1/T) gives a **straight-line plot** of the rate data (Figure 4.9).

---

[11] **Walter Hans Schottky** (1886-1976), German physicist, who discovered the ***Schottky effect*** of thermionic emission, which is the increase in electron current leaving a heated filament, when an external electric field is applied.

[12] **Yakov Ilyich Frenkel** (1894-1954), Russian physicist.

[13] **Svante August Arrhenius** (1859-1927), Swedish chemist.

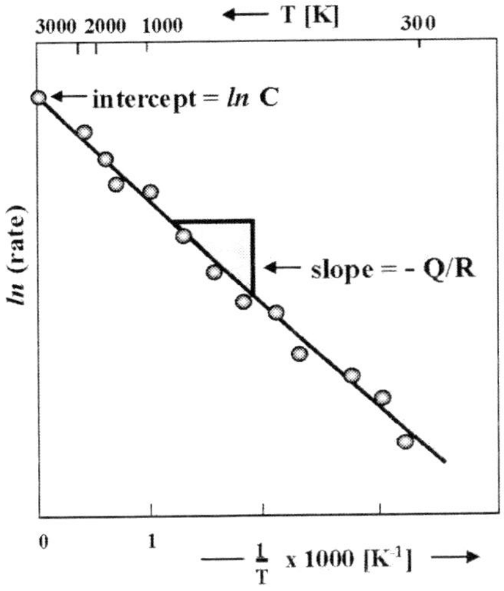

Fig. 4.9 Arrhenius plot: *ln* of process rate vs. 1/T. The slope of this curve defines the activation energy (slope = –Q/R).

The **slope** in this plot is the $-Q/R$ ratio. Extrapolation of the plot to $1/T=0$ gives the intercept equal to $\ln C$.

Commonly, the *Arrhenius plot* is used to obtain the value of the **activation energy**, Q, of a specific process. This value can indicate the specific mechanism involved that drives the process. ***Note*** that when different mechanisms are involved simultaneously, the slowest one is rate-determining (rate limiting) for the overall process.

Moreover, knowing the process rate at two temperatures allows us to *predict* the rate at a 3rd temperature. Similarly, knowing the process rate at any given temperature and Q allows the rate at any other temperature to be determined (compare Figure 4.9).

4.2.2 Activation Energy

To appreciate why the *Arrhenius plot* is so important, we have to dwell on the concept of the *activation energy*.

Dividing both Q and R by Avogadro's number, $N_A = 6.023\text{x}10^{23}$ mol^{-1}, gives:

$$process\,rate = C \cdot e^{-q/kT} \quad (4.3)$$

where $q = Q/N_A$ is the activation energy per atomic scale unit (for atoms, electrons, or ions), and $k = R/N_A$ is the Boltzman constant = 13.8x10^{-24} J/K = 8.62x10^{-5} eV/K.

If we now compare equation 4.3 with the **Maxwell[14]-Boltzman[15] distribution**, given in equation 4.4, an interesting parallel becomes obvious:

$$P \propto e^{-q/kT} \quad (4.4)$$

where P is the **probability** to find a molecule at an energy ΔE **greater** than the **average energy** typical at a particular temperature. Therefore, the activation energy Q can be visualized as the **energy barrier**, q, the system has to overcome by **thermal activation**.

As temperature increases, a larger number of atoms, electrons, or ions are available to finally **overcome** this **energy barrier**, q.

Two examples are shown in Figure 4.10 and 4.11 that illustrate the basic concept of the activation energy.

---

[14] **James Clerk Maxwell** (1831-1879), Scottish mathematician and physicist.

[15] **Ludwig Edward Boltzman** (1844-1906), Austrian physicist.

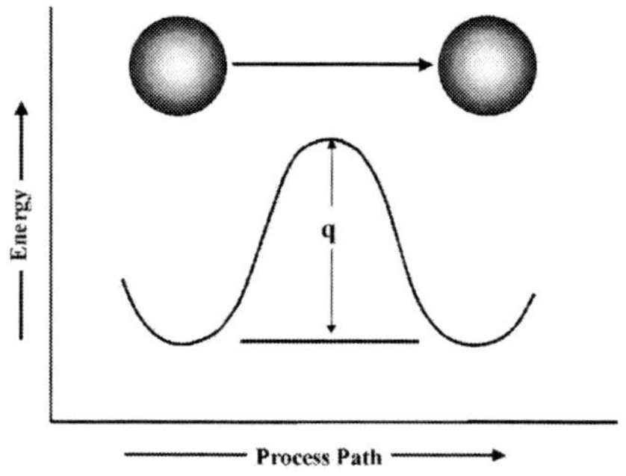

Fig. 4.10 Schematic of an energy barrier that has to be overcome, for example, by atom motion from one atomic site to another.

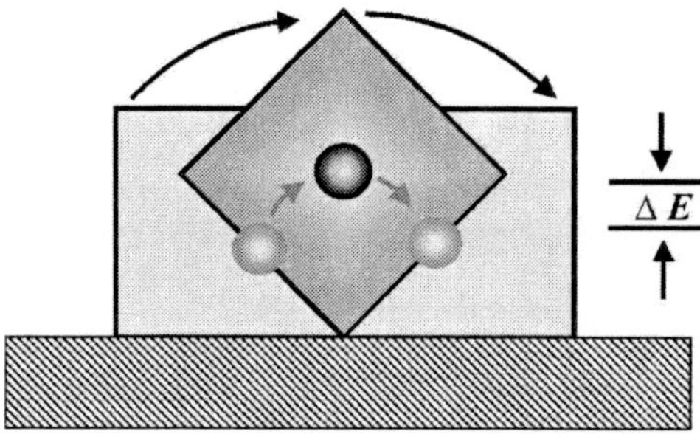

Fig. 4.11 Schematic of an energy barrier that has to be overcome during the rotation of a box (increase in potential energy).

Figure 4.10 illustrates a process path, where a single atom overcomes the energy barrier, q, to jump from one atomic site to another. Figure 4.11 shows the motion of a box (mechanical model) from one position to another. The box must overcome an increase in **potential energy**, ΔE_{pot} (center of gravity; metastable equilibrium) in order to reach a low-energy configuration again.

4.2.3 Thermal Generation of Point Defects

Formation of point defects is a direct consequence of the *periodic oscillation* of **thermal vibration** of atoms in the crystal structure. As temperature increases, the intensity of this vibration and the formation of local point defects also increases.

At a given temperature, the thermal energy of a system is fixed, but it is an **average value** that **varies** over a certain range. Hence, even at a given temperature, a certain fraction of atoms have **sufficient thermal energy** to produce point defects.

This fraction **increases exponentially** with temperature (see also *Maxwell-Boltzmann distribution*). As a consequence, the **point** defect concentration also increases exponentially with temperature:

$$\frac{n_{defect}}{n_{lattice\ sites}} = C \cdot e^{-(E_{defect})/kT} \qquad (4.5)$$

where $n_{defects} / n_{sites}$ is the ratio of point defects to the ideal crystal lattice sites, C is again a pre-exponential constant, E_{defect} is the energy needed to create a single point defect (such as a single vacancy) in the crystal structure, k stands for the Boltzmann constant, and T is the absolute temperature.

Note that the energy needed to produce a vacancy is different from the energy required to create an interstitialcy.

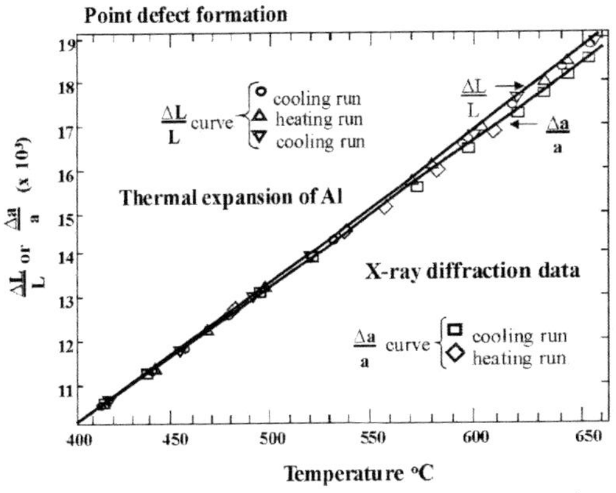

Fig. 4.12 Comparison of thermal expansion data and X-ray diffraction analysis of bulk Al metal indicating the formation of point defects at high temperature.

In the example given in Figure 4.12, **the vacancy formation** in Al metal is shown by the difference between the **thermal expansion**, ΔL/L, of the bulk metal versus the **X-ray diffraction** data that reveal the lattice plane distance, **Δa/a** (which increases linearly with temperature).

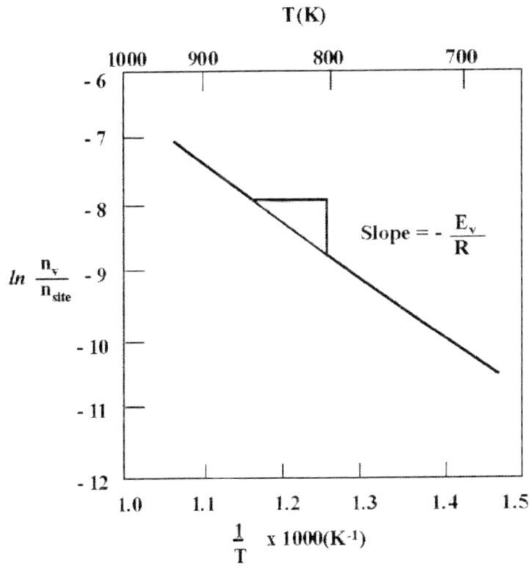

Fig. 4.13 Arrhenius plot of point defect formation rate vs. 1/T, which gives the activation energy for the formation of vacancies.

The energy term, E_V, in equation 4.6 is the energy required to create one vacancy in the Al lattice; $E_V = 0.76$ eV (energy of vacancy formation):

$$\ln \frac{n_V}{n_{lattice\ sites}} = \ln C \cdot \frac{E_V}{kT} \qquad (4.6)$$

As shown in Figure 4.13, the plot of $\ln n_V / n_{lattice\ sites}$ versus $1/T$ allows the determination of the activation energy of vacancy formation.

Example 4.1

At 400 °C the fraction of vacant Al lattice sites is 2.29×10^{-5}. What is the fraction of vacant sites at 660 °C?

First, determine C at 400 °C = 673 K:

$$\frac{n_V}{n_S} = C \cdot e^{-E_V / kT}$$

$$C = \frac{n_V}{n_S} \cdot e^{+E_V / kT}$$

$$C = 2.29 \cdot 10^{-5} \cdot e^{+0.76 eV / 86.2 \cdot 10^{-6} eV / 673 K}$$

$$C = 11.2$$

Second, calculate the vacancy fraction at 660 °C = 933 K:

$$\frac{n_V}{n_S} = 11.2 \cdot e^{-0.76 eV / 86.2 \cdot 10^{-6} eV / 933 K}$$

$$\frac{n_V}{n_S} = 8.82 \cdot 10^{-4}$$

There are about **9 vacancies** per ten thousand Al lattice sites at 660 °C.

Example 4.2

Calculate the equilibrium number of vacancies, N, per cubic meter in pure Cu at 500 °C. Assume the vacancy formation energy in pure Cu is 0.90 eV and C equals 1. The equilibrium concentration of vacancies can be expressed as:

$$n_V = CNe^{-E_V/kT} \quad (4.7)$$

with n_V being the number of vacancies per cubic meter of Cu metal, N is the total number of atomic sites per cubic meter, C is a constant (here C=1), E_V stands for the activation energy to form a vacancy in eV, k is the Boltzmann's constant (8.62×10^{-5} eV/K), and T is the temperature in K.

The number of atom sites is related to the density and the atomic mass by:

$$N = \frac{N_A \rho_{Cu}}{at.mass\,Cu} \quad (4.8)$$

with N_A being Avagadro's number (6.023×10^{23} atoms/mole), the density of Cu is $\rho_{Cu} = 8.96\ Mg/m^3$ and the atomic mass of Cu is = 63.54g/mol. It follows:

$$N = \frac{6.023 \cdot 10^{23}\,atoms}{mol} \cdot \frac{8.96 \cdot 10^6\ g/m^3}{63.54\,g/mol}$$

$$= 8.49 \cdot 10^{28}\ atoms/m^3$$

$$n_v = 8.49 \cdot 10^{28}\ atoms/m^3 \times$$

$$\left\{ \exp - \left[\frac{0.90eV}{\left(8.62 \times 10^{-5}\ eV/K\right) 773K} \right] \right\}$$

$$N = 1.15 \cdot 10^{23}\ vacancies/m^3.$$

4.3 Solid State Diffusion

At sufficiently high temperatures, atoms and molecules can be **quite mobile** in both liquids and solids. Watch a drop of ink fall into a beaker of water and see how fast the ink will spread out. This is a convincing example of liquid-state **diffusion**: mixing of materials on a molecular level.

Diffusion itself is not limited to different materials (like ink plus water). At room temperature, H_2O molecules of pure water are in continuous motion, which is called **self-diffusion.**

It is probably more difficult to visualize diffusion in **solid matter.** The primary difference between diffusion in liquids and diffusion in solids is the low rate of diffusion in solids, because solids are far more **densely packed** (close-packed crystal structures).

In fact, the energies required to drive an atom or ion through a stoichiometric crystal are so high that diffusion becomes nearly *impossible*. Therefore, **point defects** are **required** to **support diffusion.**

One has to distinguish three different **types of diffusion** processes:

- **vacancy** migration
- **interstitialcy** motion
- **interdiffusion** process.

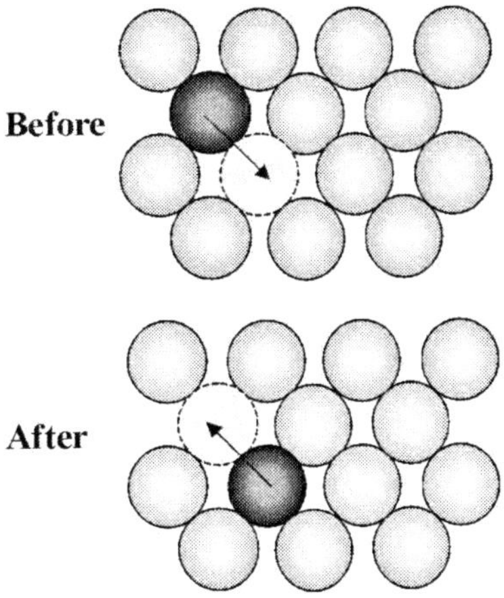

Fig. 4.14 Motion of atoms versus vacancies migration. *Note* that both species move in opposite directions.

Vacancy migration is one possible mechanism of diffusion (Figure 4.14). It is important to note that in this process the overall direction of material flow (atom or ion migration) is opposite to the direction of vacancy flow.

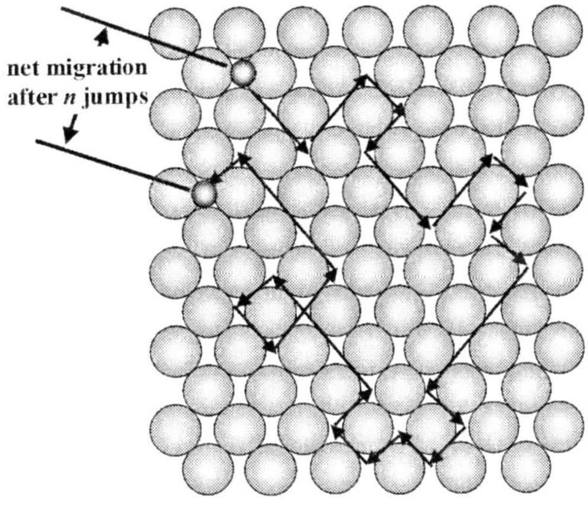

Fig. 4.15 Random motion of an intersticial carbon atom in Fe with a net migration after n-jumps.

Interstitialcy migration is a mechanism where a **random** motion of atoms on interstitial sites is involved (Figure 4.15). The randomness of the path does not preclude a **net flow** of material when there is an overall variation in chemical composition.

Interdiffusion is the net flow of material A into material B (Figure 4.16). Similarly, material B diffuses into A.

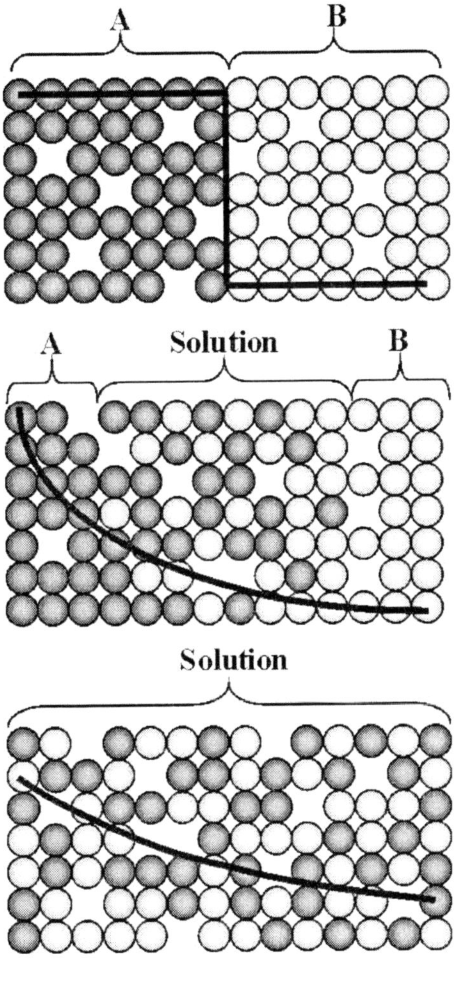

Fig. 4.16 Interdiffusion process of material A into material B and vice versa.

Note that there are also vacancies involved in this diffusion process. The formal mathematical description of such a diffusional flow of one material into another is given by **Fick's 1st law**[16].

$$J_x = -D\frac{\partial c}{\partial x} \qquad (4.9)$$

where J_x is the flux, or flow rate, of the diffusing species in the x-direction due to a given concentration gradient, $\partial c / \partial x$. The proportionality coefficient, D, is called the diffusion coefficient or simply **diffusivity**.

If we compare the graphical illustration of Fick's 1st law (Figure 4.17) with the interdiffusion process (Figure 4.16), it becomes obvious that the **concentration gradient**, $\partial c / \partial x$, at a specific point along the diffusion path **changes with time**.

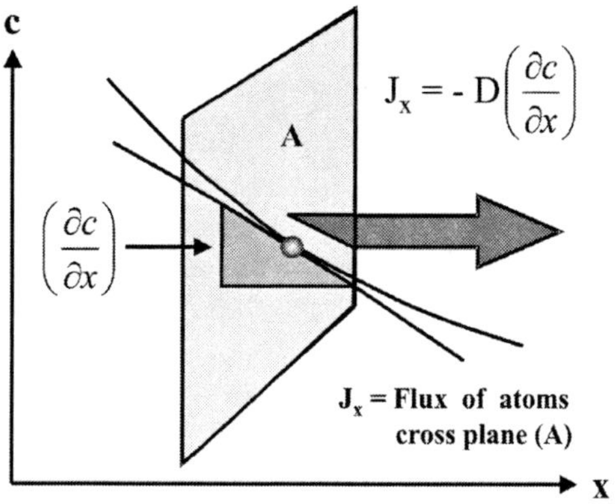

Fig. 4.17 Schematic represenation of Fick's first law of diffusion.

This **transient condition** is represented by the 2nd order differential equation also know as **Ficks 2nd law**.

$$\frac{\partial c_x}{\partial t} = \frac{\partial}{\partial x}\left(D \cdot \frac{\partial c}{\partial x}\right) \qquad (4.10)$$

For many practical purposes one can assume that D is independent of the **concentration**, c, leading to the simplified equation:

$$\frac{\partial c_x}{\partial t} = D \cdot \frac{\partial^2 c_x}{\partial x^2} \qquad (4.11)$$

which can be *illustrated* as shown in Figure 4.18). The graph reflects the *change in concentration gradient* along the x-axis (compare Figure 4.16 on interdiffusion) *with time*. This is an example of material A diffusing into a semi-infinite solid while the surface concentration of the diffusing species A, c_S, remains constant.

Technical examples of this plot would be (i) **plating** of metals or (ii) **saturation** of a metal (Al) with reactive atmospheric gas, leading to the formation of a protective Al_2O_3 surface coating.

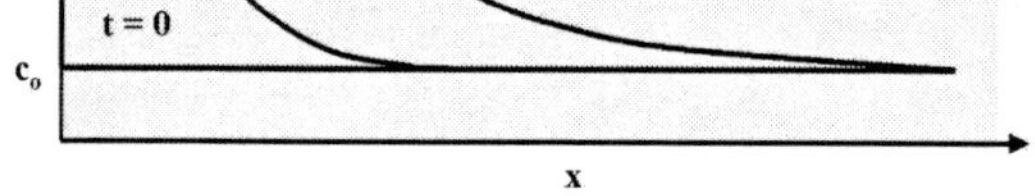

Fig. 4.18 Concentration vs. distance for different times of diffusion.

The *solution* of the *differential equation 4.8* (with the given boundary conditions) is:

$$\frac{c_x - c_0}{c_s - c_0} = 1 - erf\left(\frac{x}{2\sqrt{Dt}}\right) \qquad (4.12)$$

[16] **Adolf Eugen Fick** (1829-1901), German physiologist, who developed the 1st and 2nd law of diffusion as part of his study on blood flow.

where c_0 is the initial bulk concentration of the diffusing species. The term **erf** refers to the **Gaussian error function**[17], which is based on the integration of a *bell-shaped curve*. ***Note*** that equation 4.12 can also be expressed as:

$$\frac{c_s - c_x}{c_s - c_0} = erf\left(\frac{x}{2\sqrt{Dt}}\right) \qquad (4.13)$$

Before, we assumed a *fixed temperature*; however, there is a strong temperature dependence. Diffusivity data are one well-known example of an **Arrhenius-type equation**:

$$D = D_0 \cdot e^{-q/kT} \qquad (4.14)$$

where D_0 is the pre-exponential constant (diffusion coefficient) and q is the activation energy for defect motion. Note that here q is *not equal* to the energy of the defect, E_{defect}, but to its motion, $E_{defect\ motion}$. However, for the vacancy mechanism to be operative, vacancy formation is an integral part of the overall diffusion process, hence it follows: $q = E_V + E_{V\ motion}$.

It is common to express diffusivity data in terms of **molar quantities**; that is, with activation energy, Q, per mole of diffusing species:

$$D = D_0 \cdot e^{-Q/RT} \qquad (4.15)$$

Figure 4.19 shows various examples of diffusivity data of metals. ***Note*** that the diffusion of carbon in iron is based on an interstitial mechanism, while the diffusion of Fe in Fe is termed self-diffusion.

[17] **Johann Friedrich Gauss** (1777-1855), German mathematician.

Example 4.3

Calculate the diffusivity D for diffusion of C in γ-iron (fcc) at 927 °C. Use values of D_0 = 2.0 X 10^{-5} m^2/s, Q = 142 kJ/mol, and R = 8.314 J/(mol K). *Note* that 927 °C = 1200 K.

Using equation 4.15 it follows:

$$D = 2.0 \cdot 10^{-5}\, {m^2}/{s} \left[\exp\frac{-142{,}000 J/mol}{8.314 J/(molK) 1200K}\right]$$

and D = 1.32 X 10^{-11} m^2/s.

Example 4.4

Consider gas carburizing of a gear of 1020 steel at 927 °C. Calculate the time in minutes necessary to increase the carbon content to 0.40 % at 0.50 mm below surface. Assume the carbon content at the surface is 0.90 % and that the steel has a nominal carbon content of 0.20 %. $D_{927°C}$ = 1.28 x 10^{-11} m^2/s; use solution to Fick's second law (Gaussian error function).

C_s = 0.90 %, C_0 = 0.20 %, C_x = 0.40 %, and x = 0.50 mm; t=?

$$\frac{0.90 - 0.40}{0.90 - 0.20} = erf\left(\frac{5.0 \times 10^{-4} m}{2\sqrt{\left(1.28 \times 10^{-11}\, {m^2}/{s}\right)(t)}}\right)$$

$$\frac{0.5}{0.7} = erf\left(\frac{69.88}{\sqrt{t}}\right) \quad \text{so, erf Z} = 0.7143$$

by interpolation (corresponding Table will be given in class), one finds that Z = 0.755, which leads to:

$$\frac{69.88}{\sqrt{t}} = 0.755 \text{ and}$$

t = 8567s = 143 min.

Metals

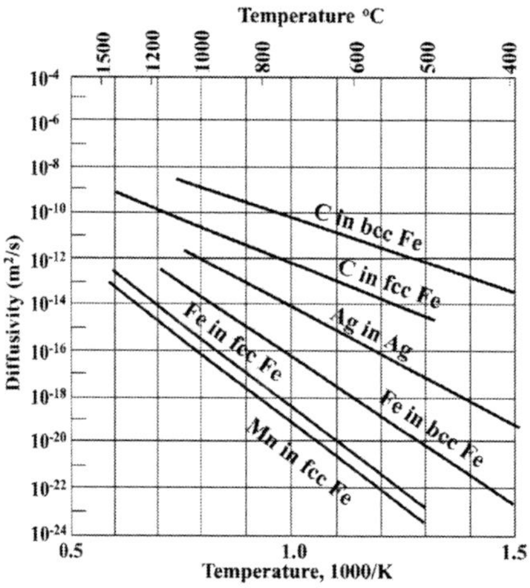

Fig. 4.19 Diffusivity data for different structures (fcc vs. bcc) and different elements diffusing in the same host material (C and Ni in fcc-Fe).

Note that self-diffusion is affected by the packing density of the particular phase (bcc vs. fcc = 0.68 vs. 0.74 APF). Hence, it follows: $Q_{bcc} < Q_{fcc} \Rightarrow D_{bcc} > D_{fcc}$.

Ceramics

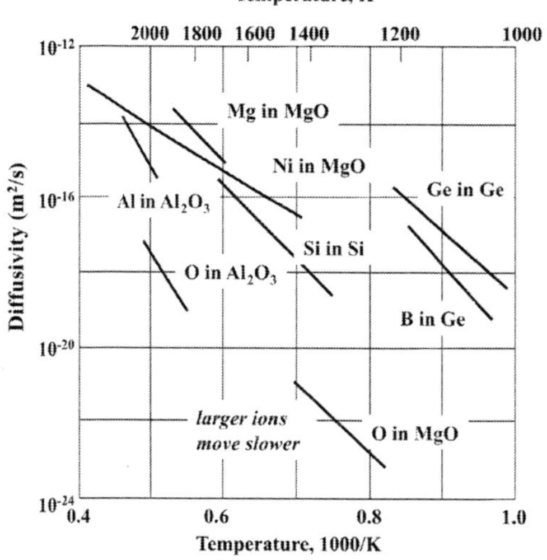

Fig. 4.20 Arrhenius plot of diffusion rate vs. 1/T for anion and cation diffusion in ceramics. *Note* that the larger ion (anion) is the slower diffusing species.

Smaller ions such as $\mathbf{Al^{3+}}$ (as compared to O^{2-}) diffuse more readily through the crystal structure (**Al_2O_3** – corundum). Similarly, the oxygen diffusion in MgO is the rate-limiting process (Figure 4.20).

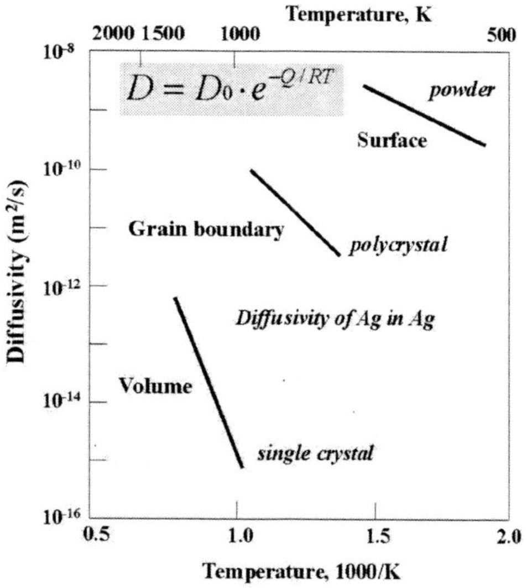

Fig. 4.21 Self-diffusivity of Ag in (i) Ag-powder, (ii) Ag-polycrystal and (iii) Ag-single crystal (surface vs. interface vs. volume diffusion).

A final word of caution: using specific diffusivity data to analyze a particular process can be misleading, since the diffusivity value strongly depends on the **route of transport** through the system (Figure 4.21) and hence the individual microstructure:

- **surface** diffusion (powder sample)
- **grain boundary** diffusion (interfaces)
- **volume** diffusion (single crystal).

4.4 Linear Defects Dislocations

While **point defects** (*zero dimensional*) are structural imperfections resulting from **thermal energy** (*thermal vibrations*); *one dimensional* defects such as **dislocations** are primarily a result of **mechanical deformation.** There are basically two major types of dislocations:

- **edge** dislocation
- **screw** dislocation.

However, it is important to note that there are also dislocations (actually the *majority*) of **mixed type** (edge plus screw components).

Dislocations are *characterized* by their:

- **Burgers vector**[18]
- **dislocation line**
- **slip plane.**

In a **perfect crystal**, an $m \times n$ atomic step *loop closes* at the starting point. However, in the region of the **dislocation**, this loop does not close. The closure vector $\vec{b}$ represents the magnitude (fraction of lattice vector) of the structural defect and is called the **Burgers vector**.

The **dislocation line** describes the path of the defect. In an edge dislocation, the dislocation line runs *along the edge* of the dislocation itself (inside the board). An edge dislocation can be envisaged as the **insertion** of an **extra ½ plane** (see also Figure 4.22) and is described by an inverted T, $\perp$.

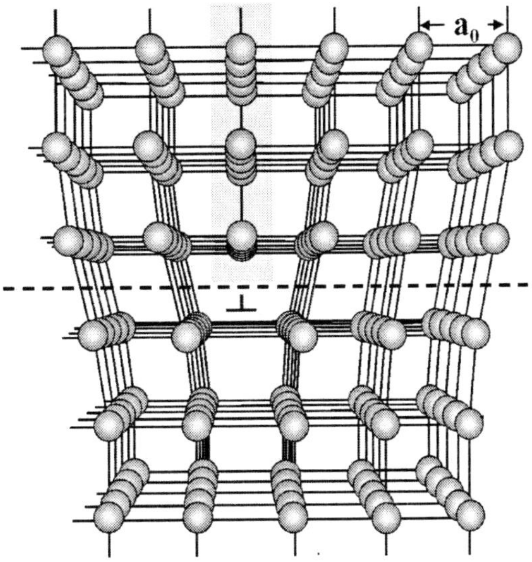

Fig. 4.22 Edge dislocation which can be viewed as the insertion of a half plane into the crystal lattice.

For an **edge dislocation**, the *dislocation line* runs **perpendicular** to the *Burgers vector*:

$$edge: \vec{b}_{edge} \perp l_{disl.}$$

Note that it is possible to create an edge dislocation by simply applying a *shear stress* to the material.

A **screw dislocation** (Figure 4.23) is similarly described as an edge dislocation. Here, however, the *dislocation line* runs **parallel** to the *Burgers vector*:

$$screw: \vec{b}_{screw} \parallel l_{disl.}$$

Screw dislocations derive their name from the **spiral stacking** of lattice planes around the dislocation line.

[18] **Johannes Martinus Burger** (1895-1981), Dutch-American fluid mechanic.

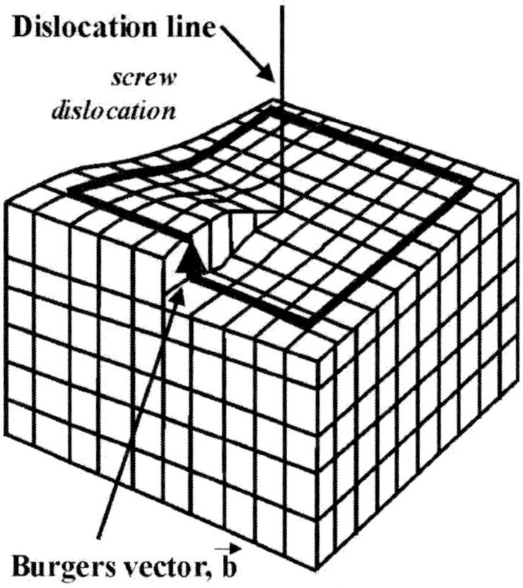

Fig. 4.23 Screw dislocation with the dislocation line running parallel to the Burgers vector.

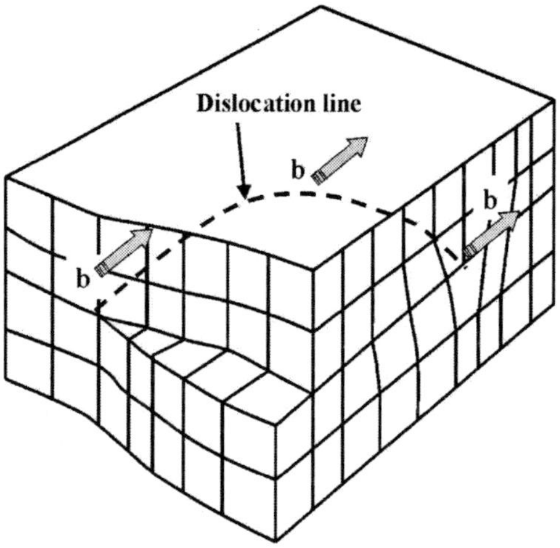

Fig. 4.24 Most materials reveal dislocations of mixed character. Here, a combination of a pure edge and a pure screw dislocation is shown. *Note* that the Burgers vector has a fixed orientation with respect to the crystal lattice.

As mentioned earlier, the **majority** of dislocations are not composed of the aforementioned pure extremes (edge or screw dislocation) but they are of **mixed character**.

The Burgers vector for a mixed dislocation type is *neither* perpendicular *nor* parallel to the *dislocation line*, but maintains a **fixed orientation** with respect to the crystal lattice (Figure 4.24), which is consistent with the previous definitions for the pure screw and pure edge dislocations.

Dislocations can be rather complicated, in particular, when the repeat distances within the crystal lattice are large. For example in Al_2O_3, the *Burgers vector* is actually **broken up** into **two** (for $\mathbf{O^{2-}}$) or **four** (for $\mathbf{Al^{3+}}$) so-called **partial dislocations** (Figure 4.25).

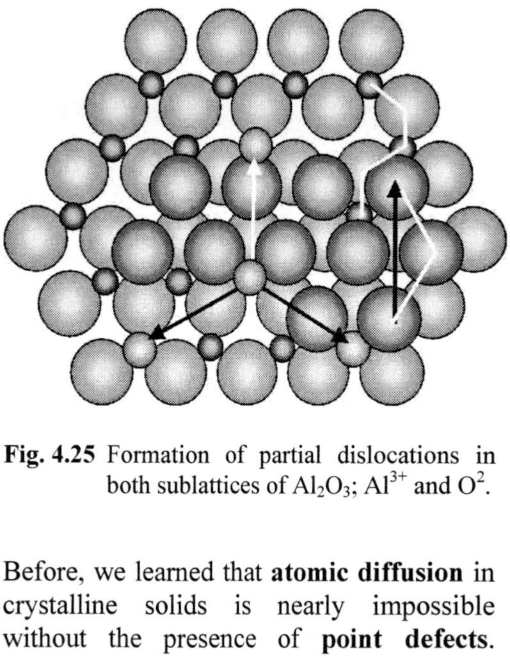

Fig. 4.25 Formation of partial dislocations in both sublattices of Al_2O_3; Al^{3+} and O^2.

Before, we learned that **atomic diffusion** in crystalline solids is nearly impossible without the presence of **point defects.** Similarly, permanent **plastic deformation** of crystalline solids is very difficult without the generation of **dislocations**.

Frenkel was first to calculate that the **theoretical** *critical shear stress* to deform a solid is about *one order of magnitude* higher that the experimentally determined bulk **shear modulus**, G. There is a correlation between **applied shear stress**, τ, and the corresponding **shear strain**, γ (for elastic, non-permanent deformation and small loads; note the correlation between Hooke's law, $\sigma = E \cdot \varepsilon$, and equation 4.16):

$$\tau = G \cdot \gamma \qquad (4.16)$$

The theoretical shear stress, τ_{theo}, for Cu is about 1000 MPa, however, the actual stress needed to plastically deform Cu is less than 100 MPa.

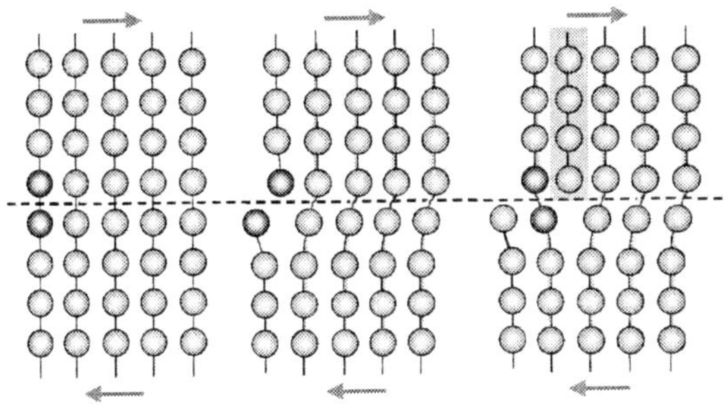

Fig. 4.26 Mechanism of dislocation motion which starts with breaking bonds and finally generates a surface step (permanent deformation).

Frenkel's observation directly relates to our daily experience. It is rather easy to open an aluminum can or to bend a thin metal sheet. However, without the presence and/or generation of dislocations, this would require a very high stress. ***Note*** that during the deformation process, the atomic planes can slide past each other (Figure 4.26).

Slip is the *motion of dislocations* that occurs in atomic **planes** with **high density** (close packed planes) and along directions with high atomic density.
The combination of *families of crystallographic plane*s with those *directions* corresponding to dislocation motion is referred to as a **slip system** (Figure 4.27).

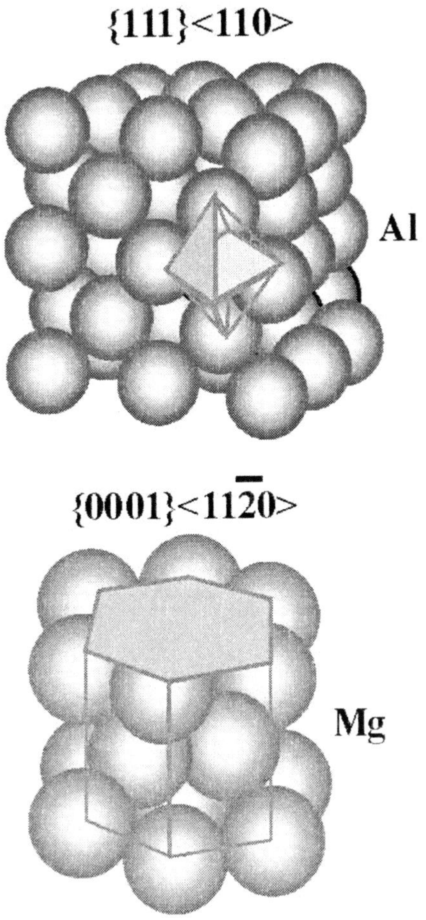

Fig. 4.27 Slip systems in fcc-Al (12 slip systems) and in hcp-Mg (3 slip systems). Note that due to the higher number of slip systems, Al is less brittle than Mg.

The difference in deformation behavior for the two metals Al and Mg (ductile versus brittle) is due to the different **number of available slip systems** (twelve vs. three).

Note that during **cold working** of a metal, deformation becomes more difficult with increasing extent of deformation. The presence and continuous formation of dislocations **hinders** the **motion** of other dislocations. Hence, cold working produces a large number of dislocations that are in fact **obstacles** to further dislocation motion (Figure 4.28). When a large number of deformation cycles are applied (remember paper clip "experiment"), the metal finally fails in a more brittle fashion.

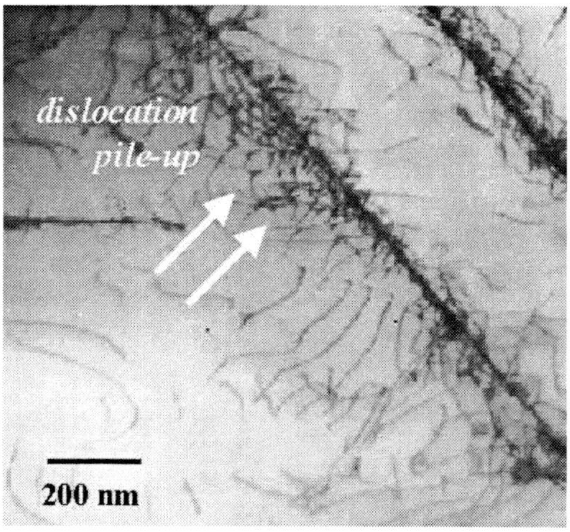

Fig. 4.28 TEM micrograph of dislocation pile-up at grain boundaries, which lowers ductility.

Similarly, **impurity** atoms or ions can also serve as *obstacles* to dislocation motion. This is the basic concept of **solution hardening** (precipitation of nm-sized secondary phase particles within the host material). Such second phase precipitation makes it harder to deform the metal.

Note that annealing at a high enough temperature (rule of thumb: 1/3 to 1/2 of the melting point) allows a metal to *heal* dislocations and thereby to reduce the dislocation density (softening).

Question

How does the presence/formation of dislocations affect mechanical deformation of metals?

4.4.1 Deformation

The atomistic view of local deformation or **plasticity** is when rows of atoms slide across each other (shear stress state). When a single crystal is loaded in tension, slip bands become visible even at the surface of the crystal (for example on a Cu surface at 0.9 % strain). The crystal starts to behave like a deck of cards. This particular deformation behavior is described by **Schmid's law**:

$$\tau = \cos\phi \cdot \cos\varpi \qquad (4.17)$$

where τ is the shear stress in the direction of dislocation motion (slip plane). Remember that theoretical calculations of the energy required to shear a single crystal have reported a shear stress, τ, which is about one order of magnitude larger than the shear stress measured experimentally. The important question to answer is: ***What is the basic mechanism of plasticity?***

The answer to the above question (discrepancy between theory and experiment) is the **mobility of dislocations.** Dislocations are line defects within a crystal and can be composed of pure edge or screw components, or they can be of mixed character.

If one wants to *control hardness*, one has to *govern plasticity*; in other words, one has to affect the formation and especially the mobility of dislocations. When dislocation mobility is lowered, plasticity is reduced and hardness increases. This effect can be gained by precipitation of small particles within the matrix (single crystal) and is called **precipitation hardening** (strengthening), a process which is very important for all **non-ferrous alloys** (like Al-alloys used in airplane construction).

Note that different planes in different crystal structures (fcc, hcp, bcc) behave differently. In **fcc** metals (Au, Ag, Cu, Ni, Al) typically the **{111}** planes are effective slip planes, while in **hcp** structures (Be, Mg, Zn) predominantly the basal planes **{0001}** are activated. In contrast, **bcc** structures also reveal a larger number of slip systems (twelve), but here the **{110}** planes are involved, which consequently affect their mechanical response, as indicated by their yield stress level: $\sigma_{Y_{bcc}} > \sigma_{Y_{fcc}}$ (see also mechanical properties).

The basic reason why bcc structures show a larger yield stress is because of the higher number of dislocations (larger number of possible slip systems which slip in a particular direction). Dislocations tend to tangle-up and pile-up quickly so that their mobility is greatly reduced. The consequence of this is a higher yield stress (onset stress of plastic deformation) for **bcc** metals (K, α-Fe, Mo, W), as compared to fcc metals.

4.5 Planar Defects

Point defects and dislocations we find within single-crystalline materials (one large imperfect crystal). However, real materials are typically composed of a large number of small single crystals touching each other. This **contact surface** in itself is a **disruption** of the periodic crystal arrangement and therefore represents a **planar defect**, which is a *two-dimensional* imperfection.

Note that when we talk about a material composed of a large number of small crystallites, we have to consider the way they are arranged and, therefore, introduce the term **microstructure**.

We have to distinguish between different geometries of boundaries (planar defects) between different grains or at the surface:

- **surface**
- **twin** boundaries
- **grain** boundaries
- **tilt** boundaries
- **twist** boundaries
- **grain-boundary films**.

4.51 Surface

All materials exhibit a surface. It is an **abrupt termination** of the periodic structure of the crystal lattice itself. The surface atoms are somewhat different from those in the bulk (coordination number, dangling bonds).

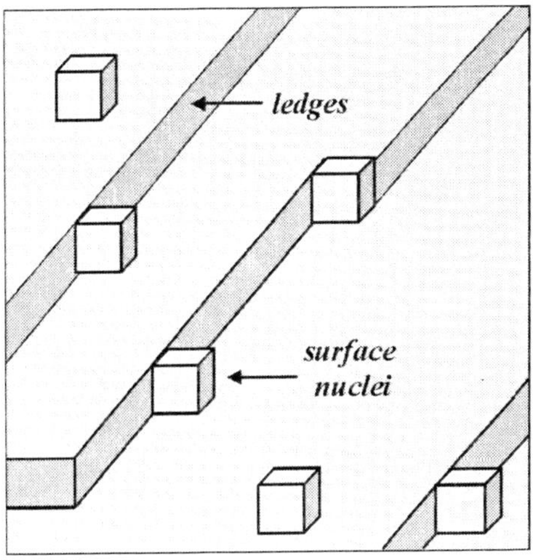

Fig. 4.29 Hirth-Pound model of the topography of a surface (formation of ledges rather than an atomically flat surface).

As a consequence, the surface atoms experience a *different bonding strength* and some *asymmetry* of geometry. In this context, note that the **Hirth-Pound model**[19] predicts **ledges** at the surface rather than atomically smooth lattice planes (Figure 4.29). Hence, all material reveal atomic steps at the surface.

4.5.2 Twin Boundaries

Twin boundaries are *highly* **symmetrical interfaces** and represent mirror images of an adjacent crystalline region (Figure 4.30).

Such a lattice **discontinuity** can be produced by simply deforming metal components, bcc and hcp, or by annealing fcc metals. Note that the formation of deformation twins can also occur in ceramics such as Al_2O_3, when local shear stresses are applied.

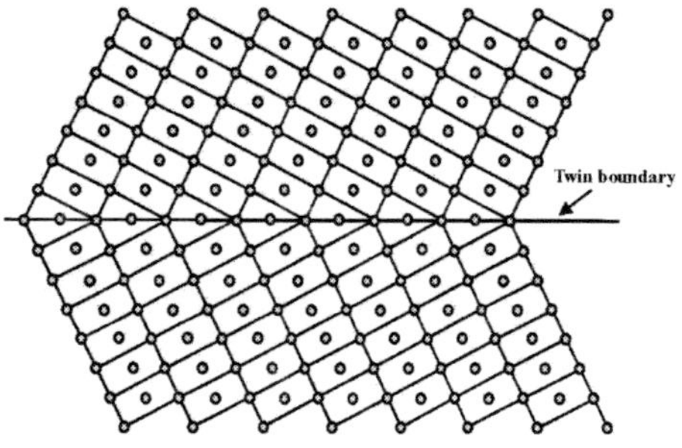

Fig. 4.30 Schematic showing the geometry of a twin boundary (mirror image).

4.5.3 Grain Boundaries

The direction and rate of growth of individual nuclei (seeds) depends on how much **free energy** is lowered during this process. The small grains grow until they **impinge** and form a discrete contact area between them, as shown in Figure 4.31.

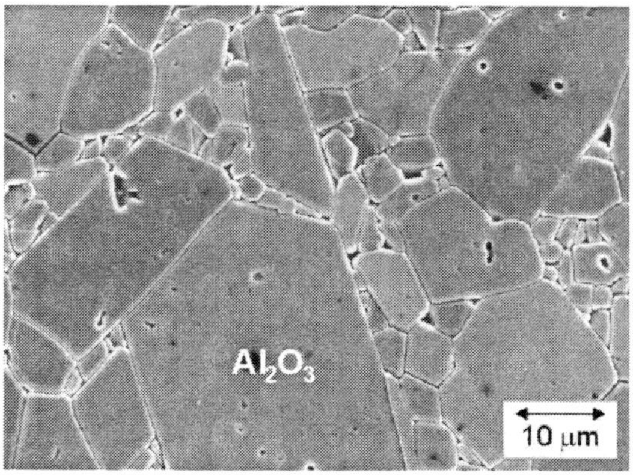

Fig. 4.31 SEM image if a polished cross section of Al_2O_3. The bright line between matrix grains indicates the interface area.

Grain Boundaries are regions between two grains with **no specific orientation relation** of the two adjacent single crystals. Such grain boundaries are a predominant feature of nearly all the materials that surround us. One interesting **exception** are **turbine blades** (for jet engines), which are in fact single crystals grown in a specific orientation in order to minimize creep deformation during operation at high temperature and high rpm.

Note that many material properties are *highly sensitive* to the specific grain structure inside the material (fine, coarse, elongated, textured etc.). Examples are transparent Al_2O_3 or the new technology of *nanomaterials*.

[19] **John Hirth** (1930-today) and **Guy Marshall Pound** (1920-1988), American metallurgists.

Tilt Boundaries are grain boundaries that are only **tilted** a few degrees relative to each other. This tilt creates a **misfit** between the two lattices that is in many cases compensated by the incorporation of pure edge dislocations (see Figure 4.32). ***Note*** that the repeat distance of such edge dislocations corresponds to the respective *lattice misfit.*

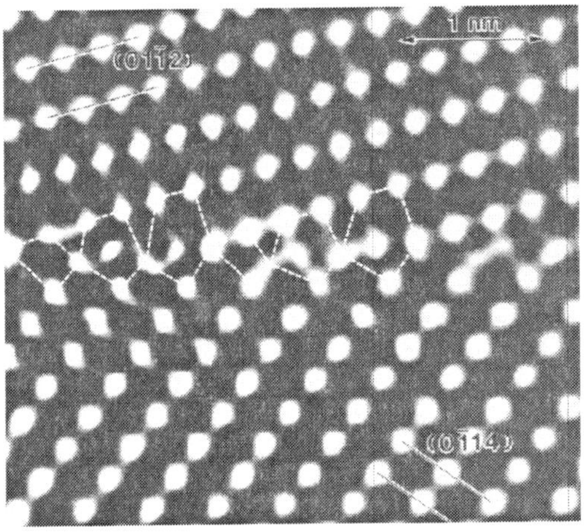

Fig. 4.32 High-resolution TEM image of a tilt boundary in Al_2O_3.

To understand the structure of tilt boundaries, the concept of the **coincident site lattice** (CSL) has been developed. The **fraction** of **coincident sites** in either lattice is represented by the symbol Σ:

$$\Sigma^{-1} = \frac{1}{5}; \quad \Sigma = 5 \ \textit{tilt boundary}$$

Depending on the **misorientation angle**, one distinguishes between **high** and **low tilt** grain boundaries.

Note that high tilt grain boundaries commonly contain so-called secondary dislocations since their Burgers vector, $\vec{b}_{gb}$, differs from the Burgers vector, $\vec{b}_b$, found in the bulk of the sample (the latter are called primary dislocations).

Twist Boundaries simply mean that the rotation axis is not parallel to both grains (zone axis), but **perpendicular** to the direction of orientation.

Grain Boundary Films are a characteristic feature first observed in non-oxide **ceramics** (SiC, Si_3N_4). The occurrence of intergranular films strongly depends on processing conditions. In the example shown in Figure 4.33, a thin (about 1.0 nm) **amorphous silicious phase** is present right at the interface between the two adjacent Si_3N_4 grains.

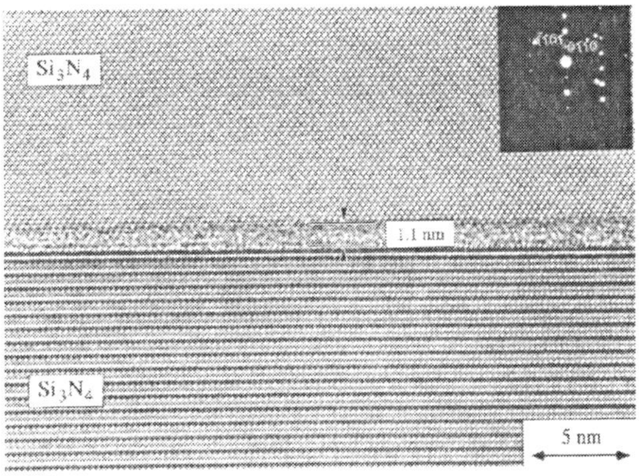

Fig. 4.33 High-resolution TEM image of an amorphous intergranular film in Si_3N_4.

Note that the presence of an amorphous secondary phase with different composition strongly affects the **high-temperature properties** of the materials. Remember: diffusion is faster in relatively wide, open structures such as glasses.

Note that interfaces are a very important **microstructural feature** of all types of materials and that they can strongly affect the resulting material properties.

5 Phase Diagrams

In this chapter we will learn about phase diagrams, which will give us an important tool to understand **microstructure-sensitive properties** which is also closely related to **microstructure evolution**. The first step is to explore the ways microstructures are developed. Phase diagrams will guide us in answering questions such as: What microstructure can we expect at a given temperature and composition depending on the experimental conditions. ***Note*** that in asking these questions, we imply that the material is at **equilibrium** condition, which is not always the case (rapid cooling).

5.1 The Phase Rule

The discussion of microstructure development via phase diagrams begins with the phase rule, which identifies the number of microscopic phases associated with a given set of **variables** (temperature, pressure, and composition) that describe the nature of the material. To quantify the volume fraction of each phase present in material, we will later make use of the **lever rule** (section 5.4).

A **phase** is a *chemically* and *structurally* **homogeneous** portion of the microstructure. A single-phase microstructure is typically polycrystalline where each grain only differs in orientation, but not in composition (e.g., pure Cu metal).

A phase must be distinguished from a **component**, which is a distinct *chemical substance* from which the phase itself is formed. For example, in the **Cu-Ni** system that forms a complete solid solution (0-100%), the alloy of a specific composition is the single phase, but **Cu** and **Ni** are the two components of this alloy. Similarly, in the case of mixing ceramic oxides such as MgO with NiO, both **MgO** and **NiO** are the two components and the formed solid solution **(Mg,Ni)O** is the phase.

Apart from a phase and a component, a **3rd term** has to be defined that is the **degree of freedom** of the system (temperature, composition, and pressure). The general relationship between microstructure and the experimental variables is given in the **Gibbs phase rule** [20]:

$$F = C - P + 2 \qquad (5.1)$$

where **F** is the degree of freedom, **C** is the number of components and **P** is the number of phases. The 2 in this equation comes from the limitation of the experimental variables to two, temperature and pressure. If for example the pressure is fixed, the above phase rule can be rewritten as:

$$F = C - P + 1 \qquad (5.2)$$

If the metal contains a single impurity, we now have two components and, therefore, solid and liquid phase can coexist over a relatively wide range of temperatures. However, since we have only *one degree of freedom*, composition will simultaneously change with a change in temperature. Therefore, composition is a so-called **dependent variable** (see also Figures 5.3 and 5.4).

[20] **Josiah Willard Gibbs** (1839-1903), American physicist.

5.2 One Component Phase Diagrams

As an example, two well known one-component phase diagrams are shown here, the phase diagram for **water** (Figure 5.1) and for **pure iron** (Figure 5.2)

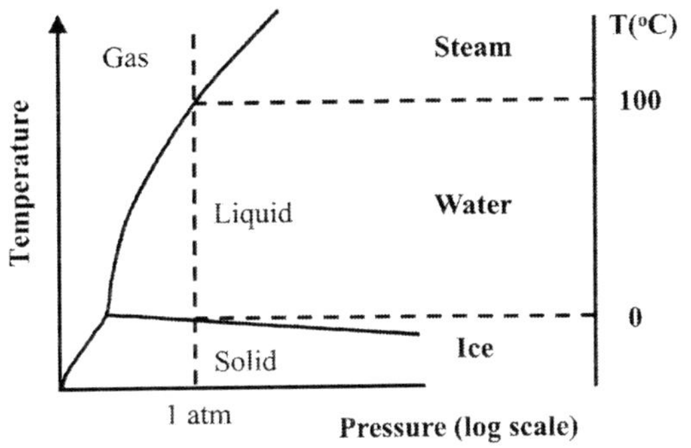

Fig. 5.1 Single component phase diagram of water.

Note that in the phase diagram of water there is one point (at low pressure) where vapor phase, ice, and water can coexist. This is termed the **invariant point.**

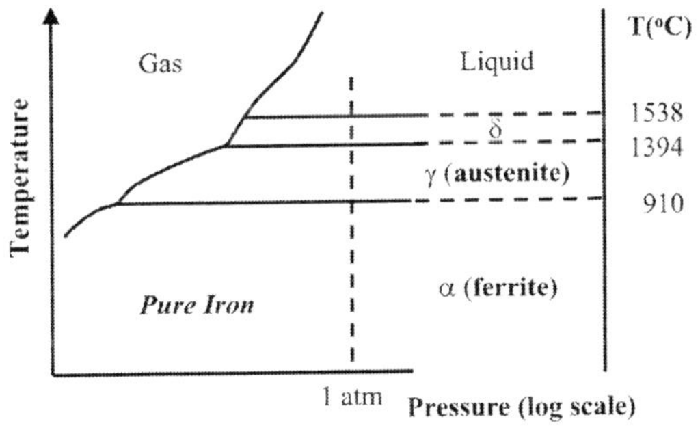

Fig. 5.2 Single component phase diagram of pure iron.

5.3 Two Component Phase Diagrams

When a system contains *two components*, its graphical representation is called a **binary phase diagram**. Note there are also *ternary* and *quarternary* phase diagrams; however, to understand the principles of phase diagrams in general, it is sufficient to limit the discussion here to binary systems.

We will distinguish between:

- **binary** phase diagrams with **complete** solid solution
- **eutectic** phase diagrams with **no** solid solution
- **eutectic** phase diagrams with **limited** solid solution
- **eutectoid** phase diagrams
- **peritectic** phase diagrams

5.3.1 Binary Phase Diagram With Complete Solid Solution

At high enough temperature, any composition will eventually have melted completely giving the **single liquid phase field, L** (see Figure 5.3).

At relatively **low temperature**, there is also a single phase field due to solid solution of the two compounds A and B; the **solid solution phase field**, ss. *Note* that it is assumed here, that the two compounds are *completely soluble* in both the liquid and the solid state.

Between these two single phase regions/fields, a **two-phase region** exists, which is commonly labeled as **L+ss**. The **upper boundary** of this two-phase region is called the **liquidus** while the **lower boundary** is denoted as the **solidus**.

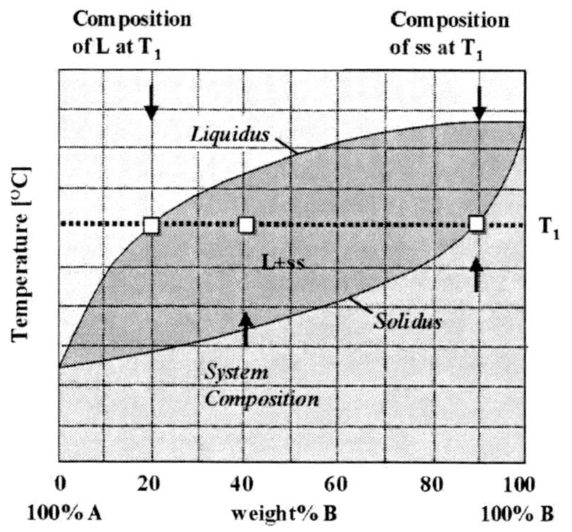

Fig. 5.3 Binary phase diagram (complete solid solution) showing the liquidus and solidus line in addition to the tie line.

At a given **state point**, which is the system composition (within the two-phase region) at a given temperature, an **A-rich liquid** is in equilibrium (coexists) with a **B-rich solid**, as shown in Figure 5.4.

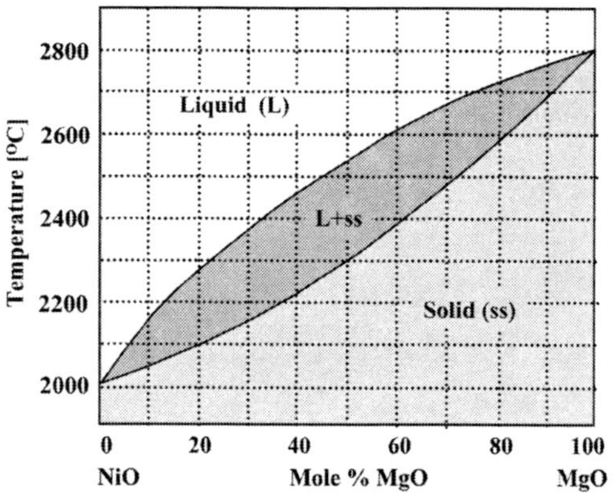

Fig. 5.4 Binary phase diagram of NiO-MgO with complete solid solution.

The **intersection** of the temperature line with both the **liquidus** and the **solidus** determines the respective **composition** of the liquid and the solid solution.

The horizontal line connecting the two points of intersection (with the liquidus and solidus line) is termed the **tie line**.

Note that there is also a close **correlation** between the **phase diagram** and the resulting **microstructure** of the material. Microstructure evolution strongly depends on the characteristics of the phase diagram and the processing conditions (heating or cooling rate, time of heat treatment at a constant temperature etc.).

5.3.2 Eutectic Phase Diagram With No Solid Solution

In this system, the **solubility** of the two components in each other is **negligible**, as for example for the **Al-Si system**, although there is in fact a small solubility of Si in Al and vice versa.

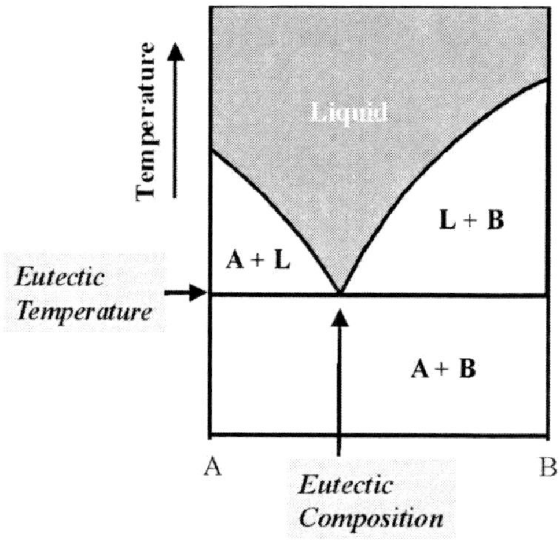

Fig. 5.5 Eutectic phase diagram with no solid solution. Note that here the two components A and B are also the two phases formed upon cooling.

There are several **distinct features** that distinguish this phase diagram from one previously discussed (complete solubility). First, at low temperature there is a two-phase field of **pure solids** A and B (no solid solution). Second, the **solidus** is a straight horizontal line that corresponds to the **eutectic temperature**. The word eutectic comes from the Greek word "**eutectos**" that means "**easily melted**". ***Note*** that the *melting temperature* of the system is *lower* than of each of the pure components.

Only at the composition of the **eutectic temperature** is the system **completely melted**. At any other composition, the material will **not fully melt** but has to be heated up to higher temperatures (crossing the 2-phase region) until it reaches the liquidus. Here we have now **two 2-phase regions**: **A+L** and **B+L**.

The liquid and L+solid microstructures are similar to those of the phase diagram with complete solid solution. The fundamental difference exists in the completely solidified microstructure. Here we find a fine-grained "**eutectic microstructure**" in which there are alternating layers of the two components, pure A and pure B.

Even during slow cooling, the system must **transform** from the liquid into the solid state rather quickly, which can be described in a way similar to a chemical reaction:

$$L \overset{cooling}{\Rightarrow} A + B \qquad (5.1)$$

Therefore, **eutectic microstructures** commonly reveal a rather **fine-grained intergrowth** of the two components (one example would be pearlite microstructure or graphic granite).

5.3.3 Eutectic Phase Diagram With Limited Solid Solution

For many binary systems, the two components are **partially soluble** in each other. The result is a phase diagram somewhat **intermediate** between the two we have just seen.

There are **two** distinguishable **ss-phases**, α and **β**, which frequently have different crystal structures. α consists of B atoms dissolved in the crystal lattice of A.

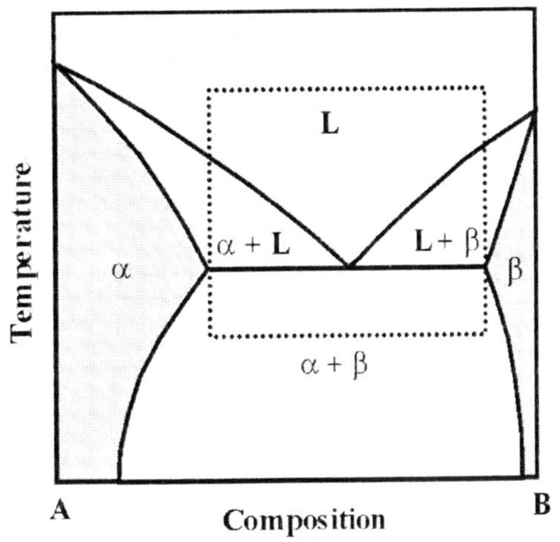

Fig. 5.6 Eutectic phase diagram with limited solid solution.

The use of the tie line to determine the compositions of a and b for a given temperature is identical as discussed for the phase diagram with complete solid solution, ss. Here, the **eutectic reaction** can be written as:

$$L \overset{cooling}{\Rightarrow} \alpha + \beta \qquad (5.2)$$

Similar to what we have discussed before, different **microstructures** evolve depending on temperature and composition.

5.3.4 Eutectoid Phase Diagram

Some of the binary systems contain an additional solid-state analog to the eutectic reaction, which is called the eutectoid reaction:

$$\gamma_{eutectoid} \overset{cooling}{\Rightarrow} \alpha + \beta \qquad (5.3)$$

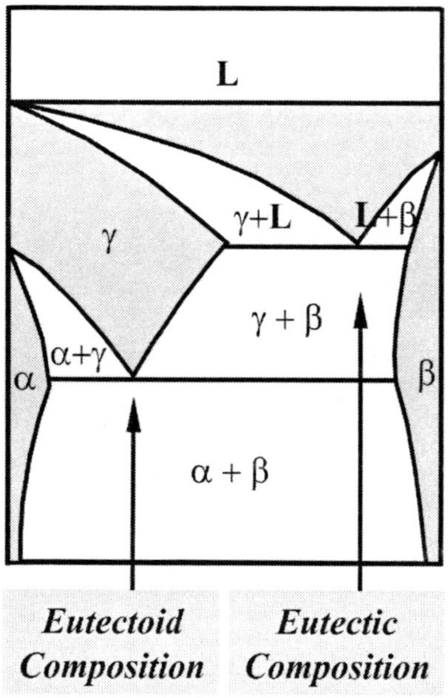

Fig. 5.7 Eutectoid phase diagram. Note that in this phase diagram both the eutectic and the eutectoid reaction can occur.

Here, similar to the eutectic system, the **eutectoid microstructures** will be rather fine-grained.

Note that this phase diagram **contains both** a **eutectic reaction** and its solid-state analog, the **eutectoid reaction**.

5.3.5 Peritectic Phase Diagram

So far we have discussed binary phase diagrams where the pure components have **distinct melting** points (for example the Cu-Ni phase diagram). In some systems, however, the components will form stable compounds that may not have a distinct melting point.

In this simplified example A and B form the compound **AB** (Figure 5.8). The melting behavior of the compound AB upon heating is named **incongruent melting** (or incongruent decomposition), which means the composition of the liquid phase formed upon melting is different from the composition of compound AB.

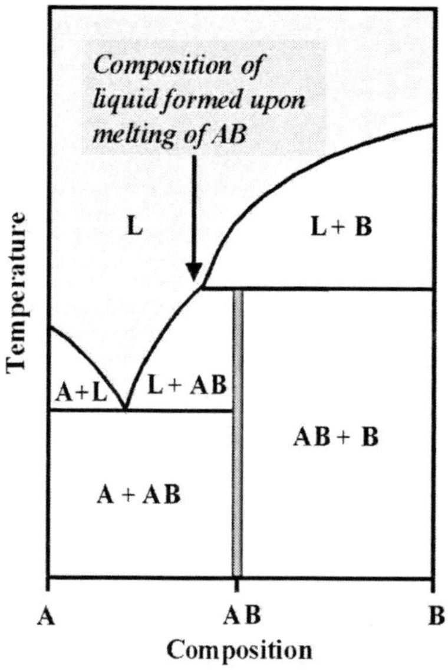

Fig. 5.8 Peritectic phase diagram. Note that the phase AB shows incongruent melting.

On the other hand, the liquid formed upon melting of components **A** and **B** will have the same composition. Such components undergo **congruent melting**.

The word "**peritectic**" comes from a Greek phrase, which means to "**melt nearby**". The **peritectic reaction** can be written as:

$$AB_{peritectic} \overset{heating}{\Rightarrow} L + B \qquad (5.4)$$

One classic example of the peritectic phase diagram is the **Al_2O_3-SiO_2 system**, where the AB compound **mullite**, $Al_6Si_2O_{13}$, is formed as an **intermediate phase** (Figure 5.9).

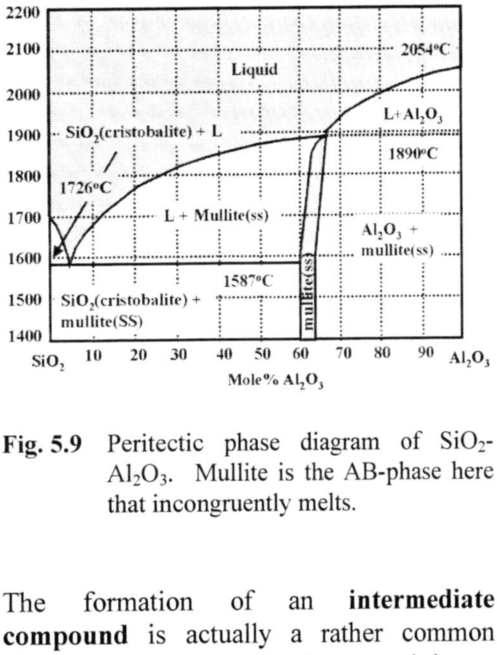

Fig. 5.9 Peritectic phase diagram of SiO_2-Al_2O_3. Mullite is the AB-phase here that incongruently melts.

The formation of an **intermediate compound** is actually a rather common feature in phase diagrams, however, it is **not necessarily** associated with the **peritectic reaction**. Figure 5.10 shows a system A-B, which also forms an intermediate compound **AB** that melts **congruently**.

Note that this phase diagram can be treated as **two similar** eutectic phase diagrams (with *no solid solution*).

The **MgO-Al_2O_3** system is also an important ceramic phase diagram that contains the intermediate phase **spinel**, $MgAl_2O_4$, which, in contrast to mullite, melts **congruently** (discrete melting point).

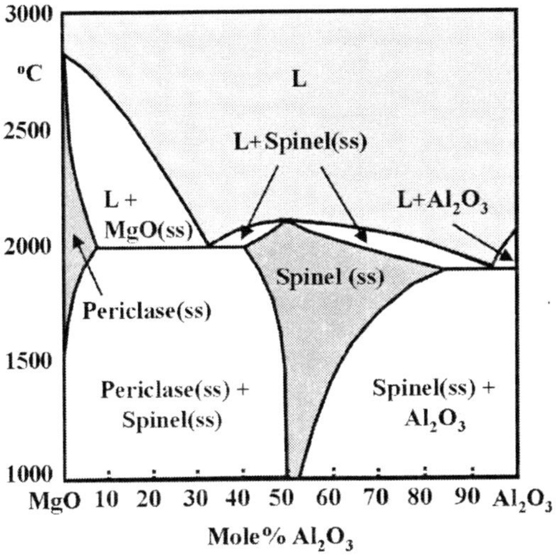

Fig. 5.10 Peritectic phase diagram of MgO-Al_2O_3. The intermediate (AB) compound in this system is spinel, $MgAl_2O_4$.

5.4 The Lever Rule

Before we have learned what phases can be expected to form and what **kind of microstructure** might form. We know already that the **tie line** determines the **composition** of each phase in a two-phase region.

In a **single-phase** region, the microstructure is always composed of **100 %** of the single phase. However, in a **two-phase** region, the quantitative determination of the amount of each phase is not that trivial.

Here we consider a **binary** phase diagram with *complete solid solution*. An **overall mass balance** requires that the sum of the two phases **equal the total** system. Assuming the total mass to be **100 g**, it follows:

$$m_L + m_{ss} = 100g$$

One could do a similar mass balance for either two components: the mass of **B** in the **liquid plus** the mass of B in the **solid solution** must equal the total amount of B:

$$m_{B_{ss}} + m_{B_L} = m_{B_{total}}$$

in our example, it would be:

$$0.80 \cdot m_{B_{ss}} + 0.30 \cdot m_{B_L} = 0.50 m_{B_{total}}$$

a mass balance in general terms is:

$$x_\alpha \cdot m_\alpha + x_\beta \cdot m_\beta = x\left(m_\alpha + m_\beta\right)$$

where x_α and x_β are the composition of the two phases and x represents the overall composition.

This equation can be **rearranged** which finally gives the **relative amount** of each phase **in terms of composition** (see also Figure 5.11):

$$\frac{m_\alpha}{m_\alpha + m_\beta} = \frac{x_\beta - x}{x_b - x_\alpha};$$

$$\frac{m_\beta}{m_\alpha + m_\beta} = \frac{x - x_\alpha}{x_b - x_\alpha}$$

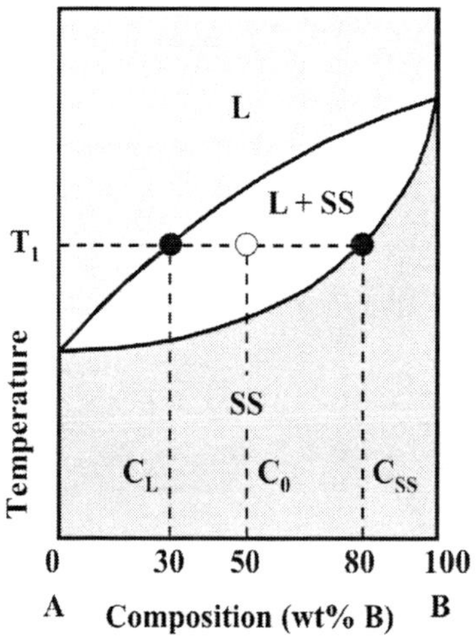

Fig. 5.11 The lever rule is used to determine the weight fractions of the different phases present (liquid vs. crystalline phase).

The lever rule is a **mechanical analogy** to the mass balance calculation, as shown in Figure 5.12.

Note that the lever rule is easier to understand when one actually plugs in numbers for the different compositions of C_L, C_o, and C_{SS}, which are 20, 50, and 80 wt%, respectively. It follows: R = 50–30 = 20 and S = 80–50 = 30.

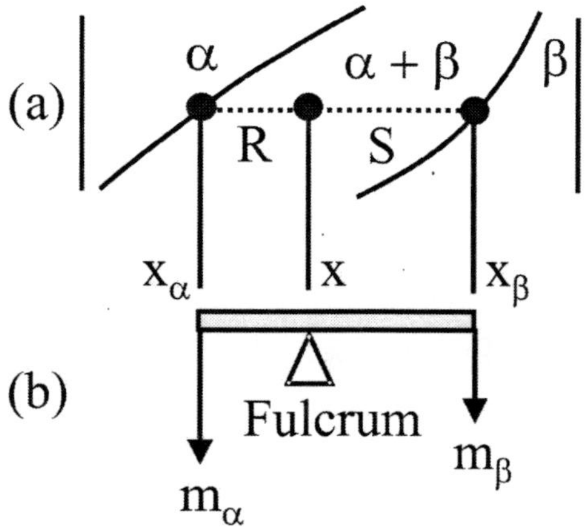

Fig. 5.12 The lever rule is a mechanical analogy to the mass balance calculation.

Example 5.1

Calculate the weight fraction of solid solution (ss) as shown in Figure 5.11. Use the equation:

$$wt\%_{ss} = \frac{C_0 - C_L}{C_{ss} - C_L} = \frac{50-30}{80-30} = \frac{20}{50} = 0.4$$

Hence, there are 40 wt% of solid solution in equilibrium with 60 wt.% liquid at temperature T_1.

The corresponding equation for the liquid is as follows:

$$wt\%_L = \frac{C_{ss} - C_0}{C_{ss} - C_L} = \frac{80-50}{80-30} = \frac{30}{50} = 0.6$$

Example 5.2

Calculate the weight fraction of liquid phase using the compositions of C_L, C_0, and C_{SS} = 32, 35, and 43 wt.%, respectively. It follows:

$$wt\%_L = \frac{C_{ss} - C_0}{C_{ss} - C_L} = \frac{43-35}{43-32} = \frac{8}{11} = 0.727$$

The corresponding weight fraction of liquid is therefore 73 wt.%.

5.5 Microstructure Development

Here we basically consider the situation of slowly cooling, which means **equilibrium** is essentially **maintained** along the cooling path.

Microstructure development of an **eutectic** composition is greatly aided by the **lever rule**, which determines both the phases that are present and their respective composition.

Microstructure evolution of a **non-eutectic** composition is far more **complex**. For example, a **hypereutectic** composition, where the starting composition is **greater** than the eutectic composition (see Figure 5.13), β_{SS} crystallites form first until the liquid reaches the eutectic composition. Final solidification occurs when the remaining (eutectic) liquid transforms suddenly to the eutectic microstructure, which is composed of fine-grained α_{SS} precipitates embedded in the β_{SS} matrix.

Hence, **large** β_{SS} grains are now embedded in the eutectic microstructure. The β_{SS} particles that precipitate during slow cooling through the *two-phase field* (L+ β_{SS}) are termed **proeutectic** β_{SS}, that is, they appear "**before**" the eutectic temperature.

Note that the fine α_{SS} crystallites in the commonly lamellar eutectic microstructure are appropriately termed eutectic α_{SS}. Precipitation of α_{SS} can occur at either:

- **triple grain junctions**
- along **grain boundaries** *or*
- inside large **matrix grains**.

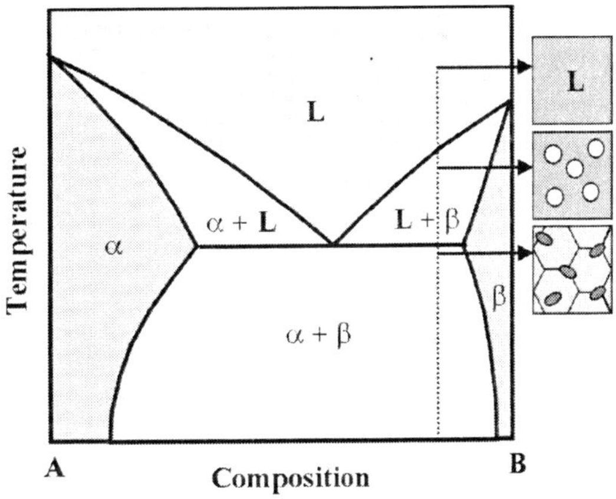

Fig. 5.13 Microstructure development of a hypereutectic starting composition.

If the starting composition is **less** than the eutectic composition, it is termed **hypoeutectic** composition. It is important to note that such a starting composition can in fact **avoid the eutectic reaction**, since all the available liquid is consumed before the system reaches the eutectic temperature, as shown in Figure 5.14. The final microstructure is composed of only one phase; the α_{SS} solid solution.

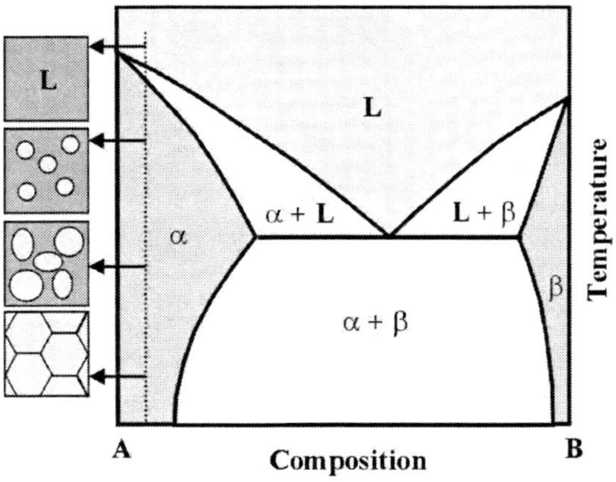

Fig. 5.14 Microstructure development of a proeutectic starting composition.

However, when the starting composition contains more of the B component (see Figure 5.15), although still being a proeutectic composition, a very different microstructure will evolve.

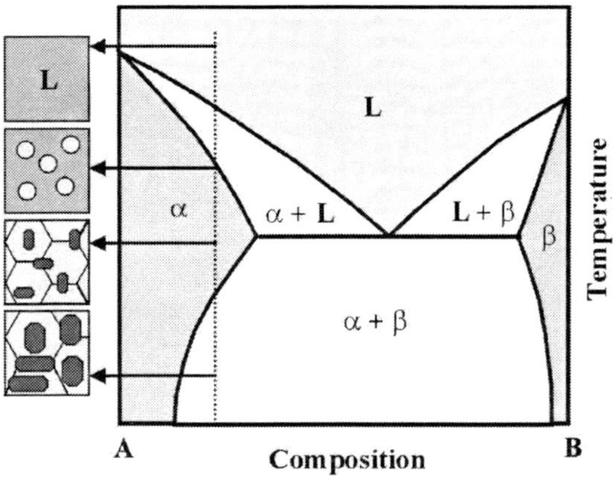

Fig. 5.15 Microstructure development of a B-rich proeutectic starting composition.

Here, **β_{ss} precipitates** will start to form when the temperature drops below the eutectic temperature. Similarly as in the case of a hypereutectic composition, precipitation of β_{ss} crystallites can occur either at triple grain junctions, along grain boundaries, or inside the larger α_{ss} matrix grains.

6 Mechanical Properties

When thinking about engineering materials, immediately a number of basic questions come to mind:

- How strong is the material?
- How much does the sample deform under a given load?
- What are the material properties after deformation?
- Is the material brittle at low temperature?
- How does the material behave at high service temperature?
- Does the material stand up to repeated loading and unloading (fatigue)?

In this chapter we explore some of the major mechanical properties of materials such as: (i) ***stress***, (ii) ***strain***, (iii) ***hardness,*** (iv) ***toughness***, (v) ***fatigue*** and (vi) ***creep***. Many of the important mechanical properties determined for metals also apply to ceramics, although the specific numbers may differ considerably.

For example, brittle fracture and creep play important roles in the structural applications of *ceramics*. In addition, in recent years *polymer engineering* succeeded in producing polymers with sufficient *strength* and *stiffness* to be able to substitute for traditional structural metals in cars. Polymers exhibit a mechanical response which is associated with their internal structure (interconnected chains) that in turn governs their *viscoelastic* deformation behavior.

The major driving forces to cope with mechanical properties in general is to try to understand *materials behavior*. One example is the optimization of the forming processes commonly used in industry for the mass production of *Al-beverage cans*.

Such products are processed at *high speed*; however, in order to maximize production speed, one needs to understand the *deformation behavior* of aluminum to generate a homogeneous wall thickness of the Al-can.

6.1 Stress-Strain Relation

The basic mechanical description of a material is given by the **tensile test** (pull test). The load necessary to produce a given elongation is monitored, as the specimen is loaded in tension at a **constant rate**. This experimental setup gives a load vs. elongation curve, as shown in Figure 6.1.

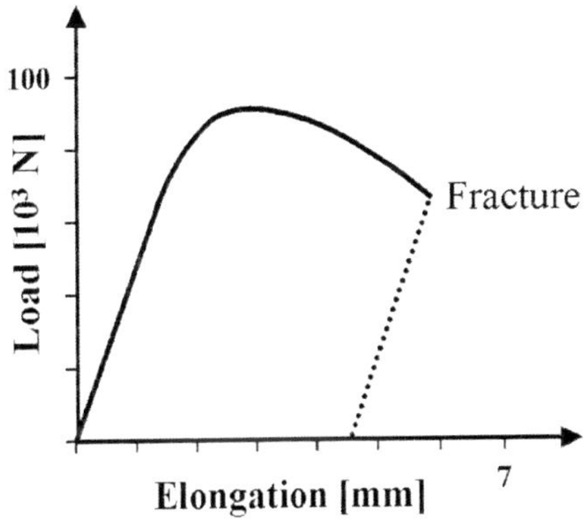

Fig. 6.1 Typical load vs. elongation curve of a tensile test (of a metal).

The **engineering strain** is defined as to how a material changes its shape when a force is applied:

$$\varepsilon_{eng} = \frac{\Delta L}{L_0}; \ \Delta L = L - L_0 \qquad (6.1)$$

The **engineering stress** is defined as the applied stress normalized by the area A_0 before deformation:

$$\sigma_{eng} = \frac{F}{A_0} \quad (6.2)$$

Commonly, one interprets the experimental load-elongation relation of a material by analyzing a plot of **engineering stress** *versus* **engineering strain,** as shown in Figure 6.2.

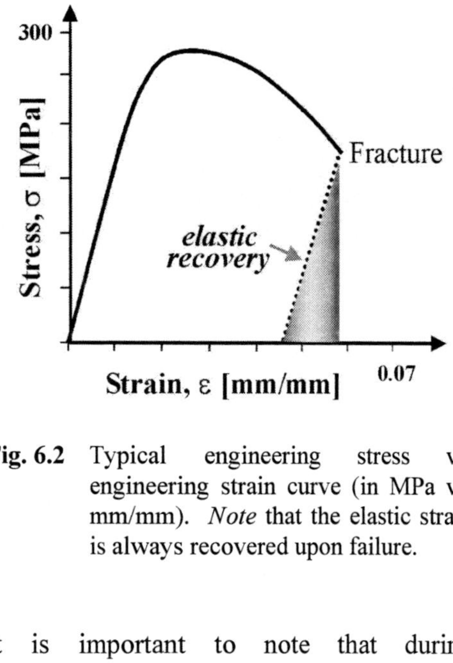

Fig. 6.2 Typical engineering stress vs. engineering strain curve (in MPa vs. mm/mm). *Note* that the elastic strain is always recovered upon failure.

It is important to note that during deformation **volume is conserved**, which results in the fact that the sample becomes narrower upon elongation (Figure 6.3).

The actual experimental measurement of the applied force/load versus elongation is converted into the corresponding **stress–strain curve** (normalized by A_0) as given above. One question that arises is: ***what to do with such a curve?***

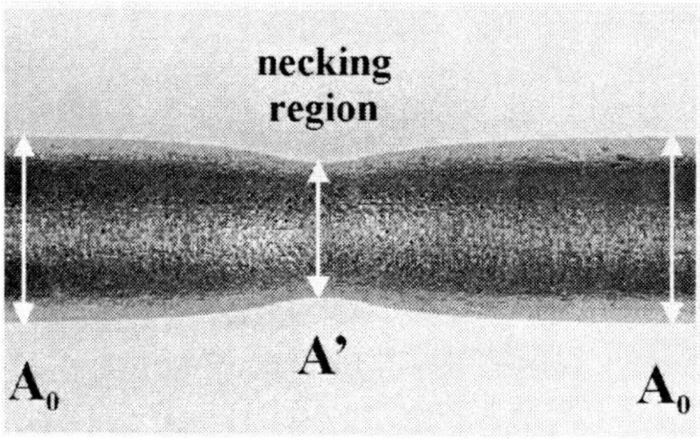

Fig. 6.3 Tensile test specimen (steel) showing the necking region in the center (conservation of volume).

One important implication of the stress-strain relation is the **stiffness**, E, of a material, also referred to as **Young's modulus**[21], which only holds for the linear (elastic) part of the stress–strain curve (see also Figure 6.6).

$$E = \frac{\sigma_{eng}}{\varepsilon_{eng}} \quad (6.3)$$

The *linearity* of the stress–strain curve in the **elastic region** is a *graphical representation* of **Hooke's law.**[22]

$$\sigma = \varepsilon \cdot E \quad (6.4)$$

Note that **elastic deformation** is ***temporary*** deformation, while **plastic deformation** is ***permanent***. The latter is *not recovered* when the load is removed. However, the small elastic strain is *always* recovered upon failure of the sample, which can also be viewed as the atoms being restored to their favored (equilibrium) position (Figure 6.2).

The plastic region is the *non-linear* portion of the stress–strain curve, generated once the

---

[21] **Thomas Young** (1773-1829), English physicist.

[22] **Robert Hooke** (1635-1703), English physicist.

total strain exceeds its elastic limit, which is termed **yield stress**, σ_Y.

It is often rather difficult to specify precisely the point at which the stress–strain curve *deviates* from linearity and enters the plastic region. Commonly two people obtain two different results. Therefore, the **0.2% offset yield stress** is defined, as shown in Figure 6.4. The usual convention is to define this offset point as the actual yield stress of the metal. Note that the term *yield strength* is also commonly used.

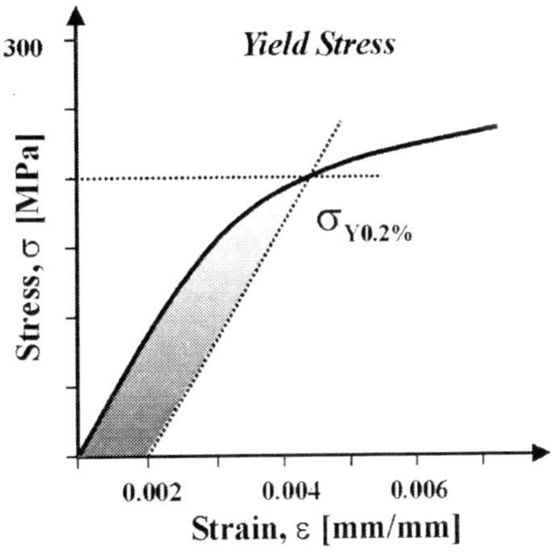

Fig. 6.4 Convention used to determine the yield stress (yield strength).

Hence, the *yield stress* is the **intersection** of the deformation curve with a **straight line** parallel to the elastic portion at 0.2 % on the strain axis. As a result, the yield strength represents the stress necessary to generate 0.2 % of permanent deformation.
Note that strain is *dimensionless* (0.002 [mm/mm]) which, when multiplied by 100 %, equals to 0.2 % strain.

The yield stress can also be seen as the "**point of no return**" where we leave the elastic deformation regime of the stress–strain curve and permanent deformation will occur (similar to bending a paper clip, which upon plastic deformation stays bent).

Before listing other mechanical properties one can obtain from a tensile test, it is worth looking at the *atomic-scale mechanism* operating in the elastic region of the stress–strain curve. The fractional deformation of the material in the initial elastic region is small so that, on the **atomic scale**, we deal with a rather small portion of the *force versus atomic separation curve*. The actual region of interest is in the immediate vicinity of the equilibrium separation distance, a_0, which corresponds to zero externally applied force, $F = 0$ (Figure 6.5). The nearly *straight-line* plot of bonding force, F, versus separation distance, a, in that region implies that there is **similar elastic response** of the material when it is stressed in **tension** or in **compression**. Experimental results indicate that this is indeed true. It should be noted that the slope of the curve shown in Figure 6.5 is related to the Young's modulus, *E*, of the material.

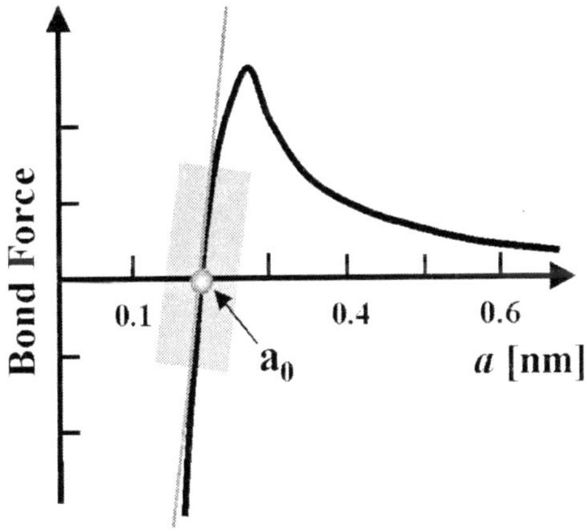

Fig. 6.5 Bond force vs. separation distance curve. The shaded area indicates the region of similar elastic response of the material in compression and tension (atomic scale).

6.2 True Stress vs. True Strain

So far we have dealt with the *engineering stress* and *engineering strain*. Now we want to extend this concept to the **true stress**, σ_{true}, and **true strain**, ε_{true}, which allows us to understand mechanical behavior from a more microstructural point of view (plasticity of the metal).

Once a load is put onto a material (threaded cylinder) and the sample rod is pulled, the material will first *stretch elastically* up to the yield point, σ_y. When further increasing the load, **irreversible deformation** (plasticity) will occur. At σ_{UTS}, the point of instability, formation of a neck (*necking*) in the center of the stretched material can be observed (Figure 6.3). Figure 6.6 shows a comparison between a stress–strain curve and a plot of true stress vs. true strain.

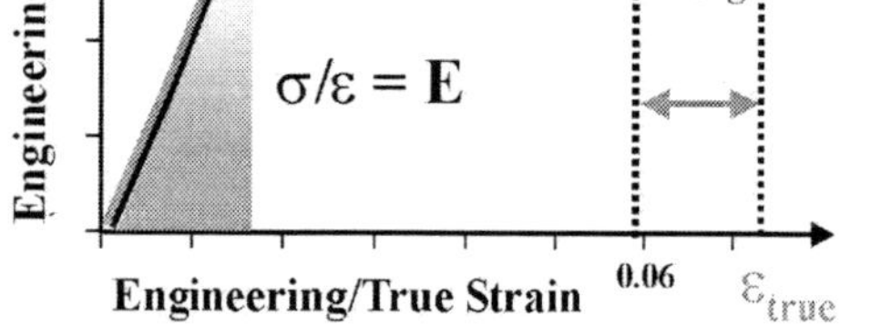

Fig. 6.6 Comparison between an engineering stress–strain curve (tensile test) and a plot of true stress vs. true strain.

Here, in contrast to the engineering stress–strain curve, the actual **true area**, A', is used in calculating the stress. This area strongly depends on the specific necking process and ***changes continuously*** throughout the experiment. The true stress is defined as:

$$\sigma_{true} = \frac{F}{A'} \tag{6.4}$$

with F being the applied force and A' the actual area. Hence, true stress is related to the real area at any given time of the experiment, which is also called the **instantaneous area**.

Similarly, the **true strain** is related to the real length change at any given time of the experiment (differential strain) by:

$$d\varepsilon = \frac{dL}{L} \tag{6.5}$$

integrating equation 6.5 gives:

$$\varepsilon_{true} = \int_{L_0}^{L} \frac{dL}{L} = \ln \frac{L}{L_0} \tag{6.6}$$

with L being an **instantaneous length.**

Looking at a stress–strain curve of a metal (engineering stress/strain) it might appear that the material in fact *softens*, because the stress drops past the ultimate tensile stress. Instead, this drop in stress is simply a result of the fact that the engineering stress and strain are defined relative to the original dimension, A_0, of the sample. Unlike the engineering stress–strain curve, the stress level of the true stress–strain curve continues to rise until the metal finally fails (Figure 6.6).

When loading the material above the critical point of the **ultimate tensile stress**, σ_{UTS}, plastic deformation continuously occurs until the material fails. Upon failure, the sample springs back elastically, which means the actual **failure strain** is defined as the total strain minus the elastic contribution (*elastic recovery*), as shown in Figure 6.2.

In Figure 6.7, the engineering stress–strain curves of a metal and a ceramic are compared. ***Note*** that the slope of each curve gives the respective Young's modulus. Au (as an fcc metal) has a large failure strain while Al_2O_3 (ceramic) reveals a low failure strain. As shown in Figure 6.7, the terms σ_y and σ_{UTS} commonly do not apply for ceramic materials.

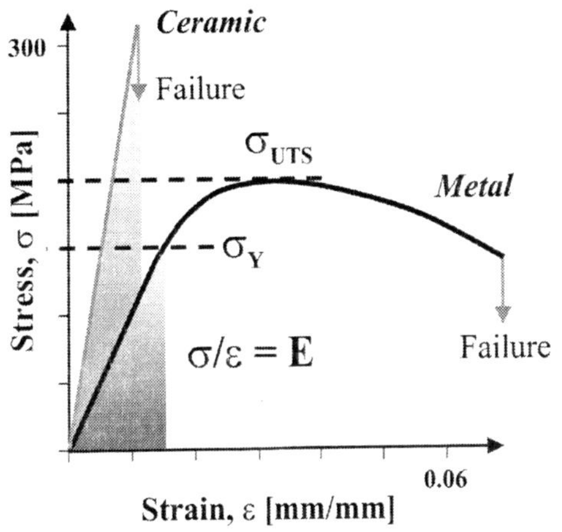

Fig. 6.7 Comparison of the engineering stress–strain curves of a ceramic and a metallic sample.

Plastic deformation is the distortion and reformation of atomic bonds leading to permanent deformation. We have already introduced the mechanism of dislocation motion as a response to externally applied load. Within the $\sigma_Y - \sigma_{UTS}$ region of the stress-strain curve, plastic deformation generates a high density of dislocations and, as a consequence, further deformation becomes more difficult. This process is called **strain hardening**. ***Note*** that strain hardening is an important factor in shaping metals via for example *cold working* (see also chapter 7.2.1).

Although intuitively one might expect that the ultimate tensile stress, σ_{UTS}, is the maximum stress the metallic sample can stand without breaking, at and above σ_{UTS} the sample begins to fail by forming (i) a neck and (ii) small pores or microcracks. Hence, in order to be able to *design* a material properly (to optimize its performance), one has to understand what happens in detail within the microstructure.

Figure 6.8 schematically shows four different stages where the microstructure is affected by the tensile test. In region I, just above σ_Y, plastic deformation occurs accompanied by dislocation formation. At σ_{UTS} (region II) necking begins. With increasing stress level, small pores are formed. In region III, those pores can coalesce forming microcracks, which then propagate until the material fails at IV. Therefore, it is safer to be more *conservative* when designing a material. Design so that the stress on the component is:

- definitely below σ_{UTS}
- preferably below σ_y, since one does not want the material to yield and initiate local stress concentrators.

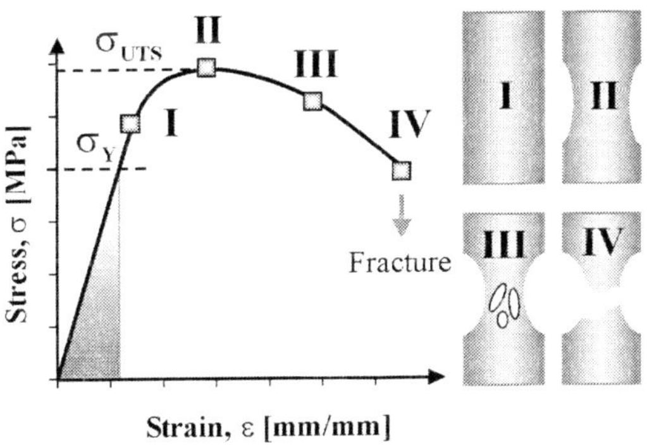

Fig. 6.8 Correlation between stress–strain curve and microstructural changes during testing.

Any metal alloy is both *strong and ductile*. A high strength alloy that in addition is rather brittle may not be a usable alloy. Similarly, an alloy that is highly deformable but has a rather low strength is not acceptable. Hence the combination of high strength with reasonable ductility is typically favored.

The general appearance of the stress–strain curve, as we have introduced before, is common for a wide range of metal alloys. However, certain alloys such as **low-carbon steels** show a distinctively different stress–strain curve, as shown in Figure 6.9.

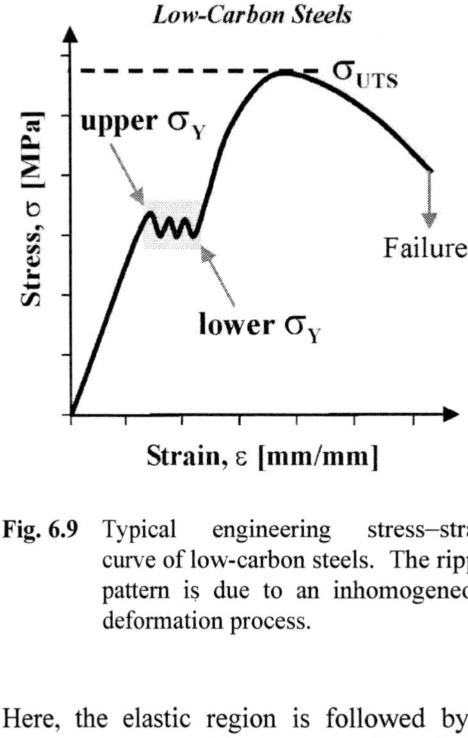

Fig. 6.9 Typical engineering stress–strain curve of low-carbon steels. The ripple pattern is due to an inhomogeneous deformation process.

Here, the elastic region is followed by a **ripple pattern**. The upper point of this pattern is termed **upper yield point**, while the point where the rippled regions ends is called the **lower yield point**. This uncommon feature is caused by an inhomogeneous deformation that begins at a point of local stress concentration.

Another very important feature of **elastic deformation** is the fact that the material **contracts** perpendicular to the extension caused by the tensile stress. This effect is characterized by the **Poisson's ratio**[23], ν:

$$\nu = -\varepsilon_x / \varepsilon_z \qquad (6.7)$$

where the stains in x and z direction are defined as shown in Figure 6.10.

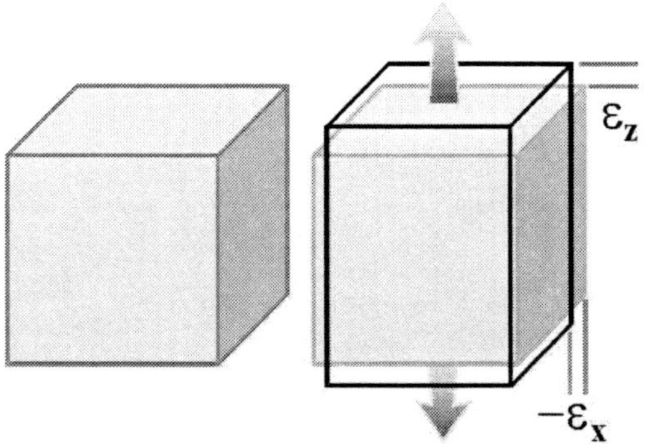

Fig. 6.10 The Poisson's ratio characterizes the contraction perpendicular to the tensile axis (elastic regime).

Although the Poisson's ratio cannot be determined by the tensile test (stress–strain curve), it is a very important and *fundamental parameter* of any elastic deformation process. Note that the Poisson's ratio values commonly fall within a *narrow band* of **0.25-0.35**. However, there are exceptions to the rule such as some polymers with $\nu \geq 0.40$.

An elastic deformation under **pure shear** loading defines the **shear stress**, τ:

$$\tau = P_S / A_S \qquad (6.8)$$

where P_S is the load on the sample and A_S is the area of the sample parallel (rather than

23 **Simeon-Denis Poisson** (1781-1840), French mathematician.

perpendicular) to the externally applied load. The shear stress produces an **angular displacement**, α, with the **shear strain**, γ, being defined as:

$$\gamma = \Delta y / z_0 = \tan \alpha \qquad (6.9)$$

as shown in Figure 6.11.

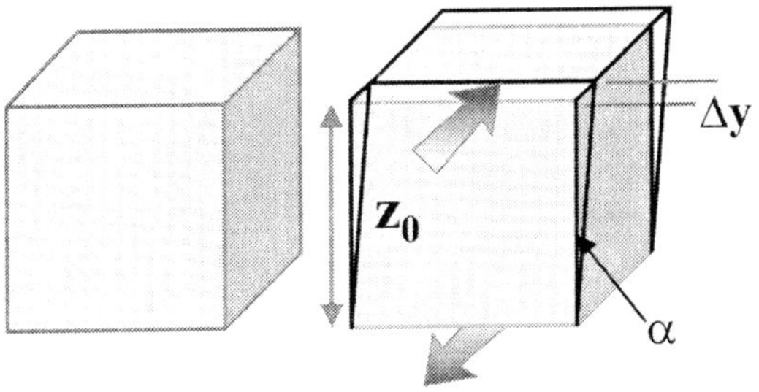

Fig. 6.11 Elastic deformation under pure shear loading.

The corresponding **shear modulus**, or modulus of rigidity, G, is defined as (compare also $E = \sigma/\varepsilon$):

$$G = \tau / \gamma \qquad (6.10)$$

The shear modulus and the elastic modulus are closely related by the Poisson's ratio (for small strains), namely by:

$$E = 2G(1+\nu) \qquad (6.11).$$

Note that, since the Poisson's ratio falls into a rather narrow band, the ratio of G/E is rather fixed for most alloys at a value of 0.4.

6.3 Hardness

The hardness test is a relatively simple alternative to the tensile test we introduced in the previous section. Hardness is defined as the *resistance* of a material *to plastic deformation*, which is related to the corresponding yield stress (yield strength) of the material. ***Note*** that since the indentation of the material produces a lot of plastic deformation, it is better to correlate **hardness** with **tensile strength**. The indenter can be either rounded or pointed and is typically made of a material much harder that the test sample (hardened steel, WC, diamond single crystal).

Hardness measurements may, for example, be used to answer the question of how materials properties have changed upon annealing (subsequent heat treatment), which is easier and faster to obtain than a stress–strain curve.

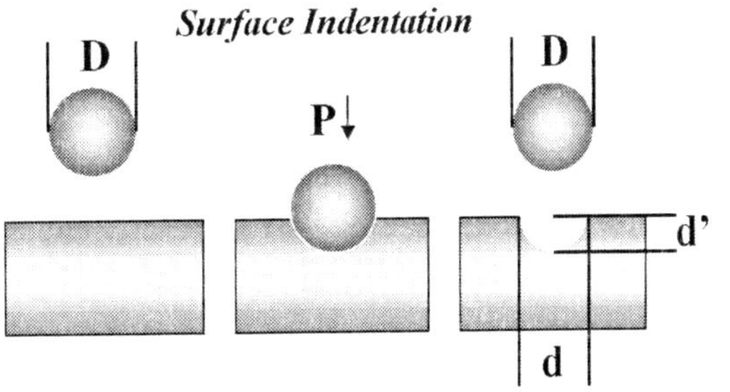

Fig. 6.12 Indentation of the polished sample surface with a sphere to measure Brinell hardness.

In principal, one could use anything to imprint the surface of a material. Commonly, a polished surface is indented (loaded) by either a ball or a pyramid and both indentation depth and width (as well as crack length) are measured to determine hardness.

Different ways have been established to measure hardness:

- **Brinell**[24] (balls of different diameter)
- **Rockwell**[25]
- **Vickers** (diamond pyramids)
- **Knoop** (pyramids).

The **Brinell hardness** number, BNH, is given by:

$$BHN = \frac{2P}{\pi D\left(D - \sqrt{D^2 - d^2}\right)} \Rightarrow \frac{load}{area} \qquad (6.12)$$

Note that the Brinell hardness number (BHN) covers a wide range of material hardness with a single scale, as opposed to the Rockwell hardness, which utilizes a number of different scales (Rockwell A, Rockwell B, etc.).

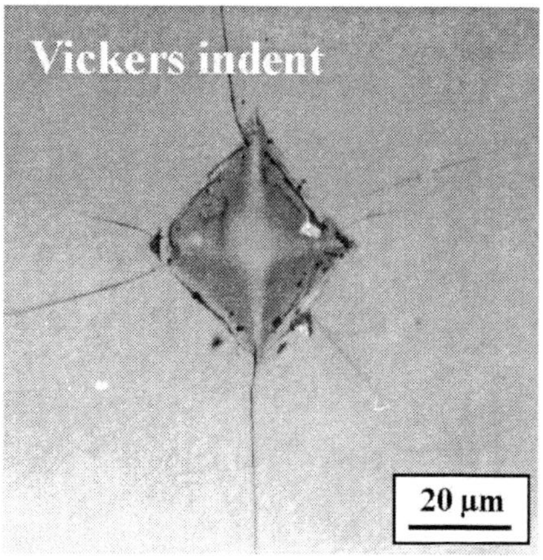

Fig. 6.13 Scanning electron microscopy (SEM) image of a Vickers indent with indent cracks on a SiCO ceramic surface.

6.4 Charpy Impact Test

The **Charpy test**[26] measures the **impact energy** necessary to fracture a pre-notched material by determining how far the hammer swings after the sample was hit and broke (Figure 6.14). This test is related to the **fracture toughness**, as **hardness** is analogous to **tensile strength**.

The Charpy test correlates with the *area under the total stress-strain curve* and the value obtained represents an energy value; ***how much energy is stored in the material during fracture***.

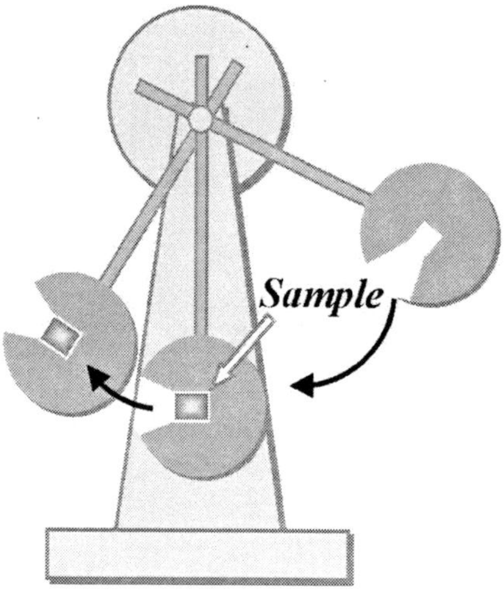

Fig. 6.14 Schematic illustrating the principle of the Sharpy test. A hammer swings and hits and breaks the sample; the end of the swing determines the impact energy.

Note that the test itself is *sensitive* to test conditions (temperature, atmosphere) as well as how sharp the notch actually is. Moreover, a very *sharp notch* creates a local *stress concentration* at the crack tip.

---

[24] **Johan August Brinell** (1849-1925), Swedish metallurgist.

[25] **Stanley P. Rockwell** invented the hardness tester in 1919, American metallurgist.

[26] **Augustin Georges Albert Sharpy** (1865-1945), French metallurgist.

6.5 Fracture

Before, we have considered material failure from a **macroscopic** picture (tensile test, hardness). Now we will consider the mechanical response of the material from a **microscopic** point of view (atomistic picture of failure).

6.5.1 Ductile-*to*-Brittle Transition

Basically we distinguish between two general kinds of fracture: **ductile** and **brittle**. Brittle fracture involves little or no plasticity (breaking a glass coffee pot), while ductile fracture involves a fair amount of plasticity (bending a paper clip).

As we already know, **fcc-Al** is highly **ductile** while **hcp-Mg** is rather **brittle**. In addition to the intrinsic crystal structure, temperature also affects plasticity in many materials. For exmple, **bcc metals** show a pronounced *temperature dependent* fracture resistance. Higher temperatures promote deformation, whereas low temperatures promote fracture. In many materials, a **ductile-to-brittle transition** can be detected when *lowering* the temperature. The main reason for this behavior is a rising yield stress with decreasing temperature. As a consequence, *tensile stresses* can cause catastrophic failure of the material **before** *shear stresses* cause deformation.

This phenomenon illustrates the essential aspect of the competition between fracture and plastic flow. If *dislocation slip* is easy, then a small crack will blunt by plastic flow at the crack tip and hence be arrested (high fracture toughness).
The basic explanation of the transition from ductile to brittle failure as the temperature is decreased is that it becomes increasingly difficult to move dislocations relative to the energy required to propagate a sharp crack, as expressed by the tensile fracture stress. With decreasing temperature, most materials exhibit an increase in the so-called **Peierls stress**[27], which corresponds to the increase in lattice friction, which in turn affects dislocation mobility.

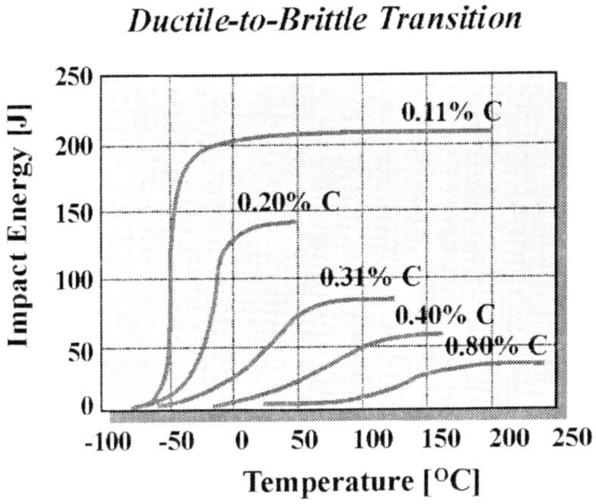

Fig. 6.15 Effect of carbon content on impact energy vs. temperature of bcc metals.

Note that the ductility of bcc metals (Fe, Mo, W) is intrinsically lower than of fcc metals. A well-known example of this brittle-to-ductile transition phenomenon in bcc metals is the failure of the Liberty ships during World War II (Figure 6.15). The entire hull of these ships cracked in half as they sailed into the cold ocean waters of the Atlantic. The mechanical behavior of the ship hull in the frigid marine environment was quite different than that in the testing laboratory (room temperature).

[27] **Rudolf Ernst Peierls** (1907-today), German-born British physicist.

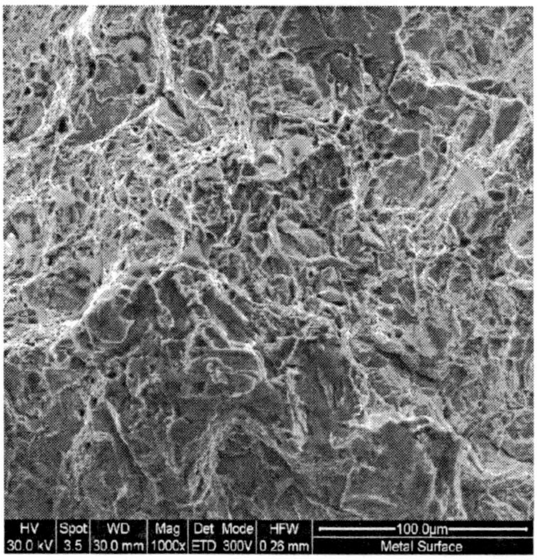

Fig. 6.16 SEM image of a typical fracture surface of a ductile metal (tensile test).

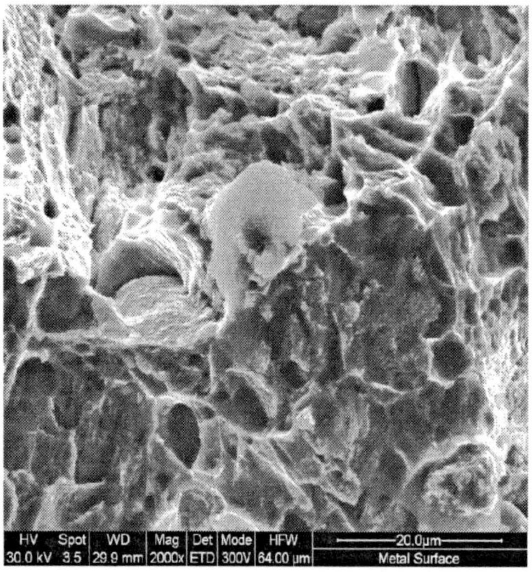

Fig. 6.17 Higher magnification SEM image of the fracture surface shown in Figure 6.16 (tensile test of a metal).

In metals it is rather difficult to drive a crack through the material, which results in a high fracture toughness. A ductile fracture surface shows characteristic features such as **shear lips**, which form a **cup-and-cone** fracture surface, as shown in Figures 6.16 and 6.17. In some cases ductile fracture surfaces also reveal some fibrous features.

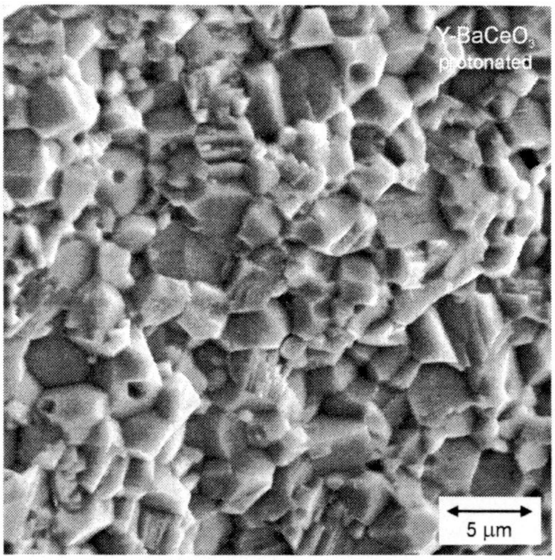

Fig. 6.18 SEM image of a characteristic fracture surface of a brittle ceramic ($BaCeO_3$).

In contrast to the characteristic features of ductile fracture as shown in Figures 6.16 and 6.17, brittle fracture is commonly characterized by a rather smooth fracture surface due to predominantly transgranular fracture (cleavage through the matrix grains). In the example shown in Figure 6.18, however, weak interface bonding leads to intergranular fracture resulting in an uneven crack path.

6.5.2 Stress Amplification

A more detailed understanding of fracture mechanics allows the prediction of how and when a crack in a given material will propagate and when failure of the component will finally occur.

One fundamental concept of fracture mechanics is the aspect of local **stress concentration** at a **preexisting flaw** (for example at a thin slit or the tip of a microcrack). Stress will be highest right at the tip of the crack. When this stress overcomes a *critical value*, the preexisting crack can *propagate* on the atomic level by simply breaking atomic bonds.

The amount of local **stress concentration** within a material that contains a small crack depends on the **radius** of the **crack tip**. The highest stress at the crack tip, σ_m, is given by the **Griffith equation**[28]:

$$\sigma_m = 2\sigma_{app}\left(\frac{c}{r}\right)^{1/2} \qquad (6.13)$$

where σ is the applied stress, c the crack length and r the radius of the crack tip. This concept was first applied by Alan Griffith in 1920, when he worked on the strength of glass fibers. Since the radius can be as small as an *interatomic spacing*, the local stress intensification can be quite large. Therefore, routinely production and handling of intrinsically brittle ceramics and glasses make the use of the Griffith equation inevitable.

The (local) stress amplification, which is also called the **stress intensity factor**, is given by:

$$K_I = Y\sigma_{app}\sqrt{\pi a} \qquad (6.14)$$

where σ is the applied stress, a stands for ½ the flaw size, and Y is a geometry factor which usually is equal to unity. At fracture, K_I equals a critical value, K_{IC} (opening mode I), and the crack propagates forward. At this stage, the applied stress equals the fracture stress, $\sigma = \sigma_f$, and the initial flaw size has a critical value, $a = a_c$, which gives:

$$K_{IC} = Y\sigma_f\sqrt{\pi a_c} \qquad (6.15)$$

When the stress level applied to a material rises, the crack can propagate a little and a increases. At a constant applied load, this will increase K_I and, when K_I equals K_{IC}, will result in **catastrophic failure** (the propagating crack cannot stop). This critical value K_{IC} is the **fracture toughness** of the material, also termed critical stress intensity factor, and most importantly is a *material property*. Typically the higher the fracture toughness of a material, the lower the yield stress/strength, σ_Y, which determines the failure mechanism: ductile vs. brittle.

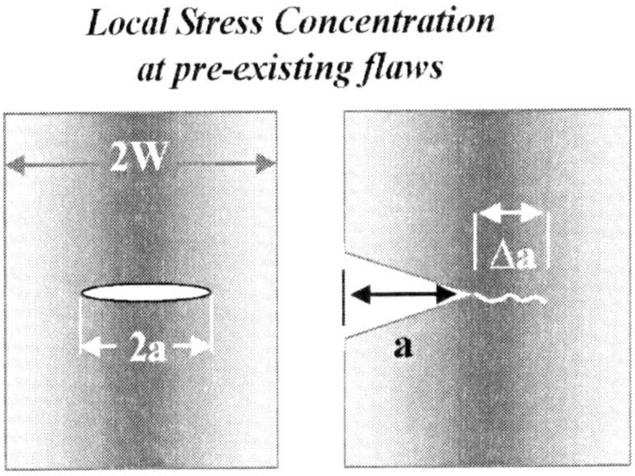

Fig. 6.19 Local stress concentration caused by intrinsic flaws; either a small slit (Y=1) or a sharp microcrack (Y=1.125). *W* is typically two orders of magnitude larger than *a*.

Note that engineers commonly do not prefer to design with ceramic components, since they are not very reliable due to their low fracture toughness. A small processing flaw within a material, with the given local stress concentration (Figure 6.19), can cause catastrophic failure of the component. Since it is rather difficult to control flaw size and flaw distribution within a component during processing, ceramics and glasses (low fracture toughness) are rather sensitive to the

[28] **Alan Arnold Griffith** (1893-1963), British engineer.

presence of pre-existing flaws and tend to fail in a more uncontrolled manner, as compared to metals or metallic alloys (see also Table 6.1).

Tab. 6.1 Fracture toughness and yield stress of different metallic alloys.

Material	K_{IC} MPa $(m)^{1/2}$	K_{IC} ksi $(In)^{1/2}$	σ_y MPa	σ_y ksi
Al-alloys				
2024-T851	26	24	455	66
7075-T651	24	22	495	72
7178-T651	23	21	570	83
Ti-alloys				
Ti-6A1-4V	55	50	1035	150
Alloy steels				
350-1	60	55	1515	220
17-7	77	70	1435	208
350-2	55	50	1550	225

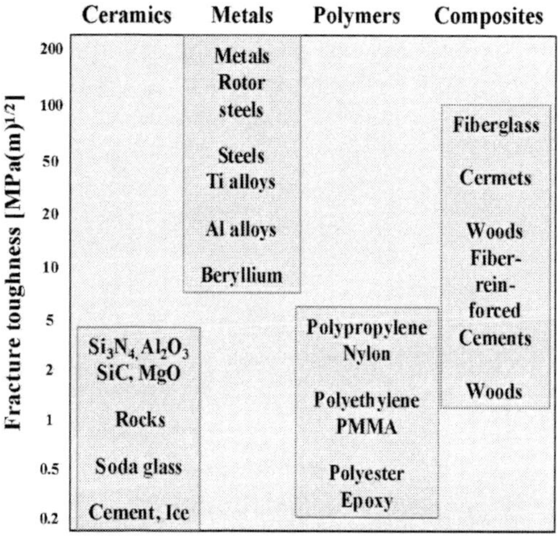

Fig. 6.20 Fracture toughness of different material groups (ceramics, metals, polymers, composites).

In order to put things into perspective (correlation between materials composition and properties), we have to look at the *fracture toughness* of different materials, which shows for example that a metal like gold (Au = low yield stress and low strength) has a high fracture toughness value, which makes *it difficult to propagate a crack.*

Example 6.1

Determine the critical crack length (in mm) for a through crack in a thick 2024-T6 Al-alloy plate which has a fracture toughness K_{IC} = 24.2 MPa·m$^{1/2}$, and which is under a load σ = 350 MPa. Assume Y = 1. Minimum crack size for catastrophic fracture/failure is then:

$$K_{IC} = \sigma_c \sqrt{\pi a}$$

$$a = \frac{1}{\pi}\left(\frac{K_{IC}}{\sigma_C}\right)^2$$

$$a = \frac{1}{\pi}\left(\frac{24.2 MPa\sqrt{m}}{350 MPa}\right)^2$$

which gives: a = 0.00152 m = 1.52 mm; however, since the intrinsic flaw size equals $2a$, the critical crack length is ⇒ 3.04 mm.

As an exercise calculate the maximum flaw size a common glass like window glass with a fracture toughness value of about 1.0 MPa(m)$^{1/2}$ can withstand under the same loading conditions (the difference in critical crack length between an Al-alloy and a glass should be about *three orders of magnitude*).

6.6 Fatigue

Up to this point, we have introduced mechanical behavior of materials under a **single load application**, either during the *slow* testing conditions of *tensile testing* or under the *fast* testing conditions of the *Charpy test* (impact energy).

There are basically three ways in which load can be applied to a test sample:

- static
- cyclic
- impact (high loading rate).

Many structural applications involve **cyclic** rather than static loading, which generates a specific condition. **Fatigue** is the general phenomenon of **material failure** after a **number of cycles** of loading (and unloading) to a stress level, which is *below* the ultimate tensile stress (Figure 6.21).

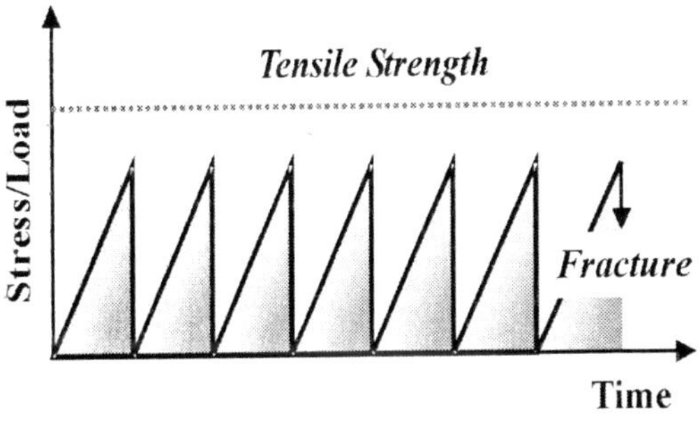

Fig. 6.21 Continuous loading and unloading of a material *below* its tensile strength limit.

The plot of stress, *S*, versus a number of cycles, *N* (often logarithmic) is called **S-N-curve**. While a material can, for example, withstand a stress level of 800 MPa during a *single loading event*, it can fracture after N=10,000 cycles at a much lower **fatigue strength** (for example 500 MPa; Figure 6.22).

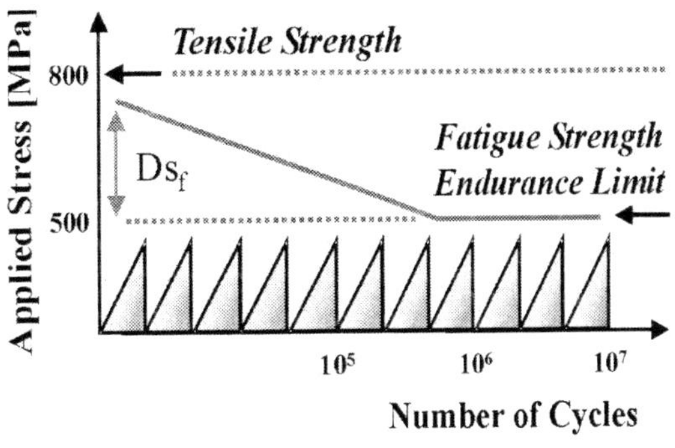

Fig. 6.22 Reduction in load bearing ability (endurance limit) with increasing number of loading/unloading cycles.

Repeated application of stress to a material creates localized deformations at the surface, which eventually will evolve into **distinct sharp discontinuities** (so-called extrusions and intrusions). Once these intrusions are formed, they tend to grow into small cracks at the surface (a small surface flaw). Such surface flaws serve as local **stress concentrators** (see stress-intensity factor). Consequently, the material has a **reduced load-carrying ability**.

Commonly, the decay in strength with increasing number of cycles reaches a limit. This **fatigue strength** or **endurance limit** is characteristic for ferrous alloys. Non-ferrous alloys tend not to have such a distinct endurance limit; however, there is still a drop in the maximum load the material can withstand after numerous loading/unloading cycles.

When we consider a **ductile** material where it is difficult to drive a crack, dislocation motion occurs during loading and unloading. Due to the high mobility of dislocations in ductile materials (metals), they tend to **pile up** at intrinsic grain boundaries (Figure 6.23; see also Figure 4.28), which generates a rather *high stress level* at the interface and can initiate the formation of microcracks resulting in a *"semi-brittle"* behavior of the metal.

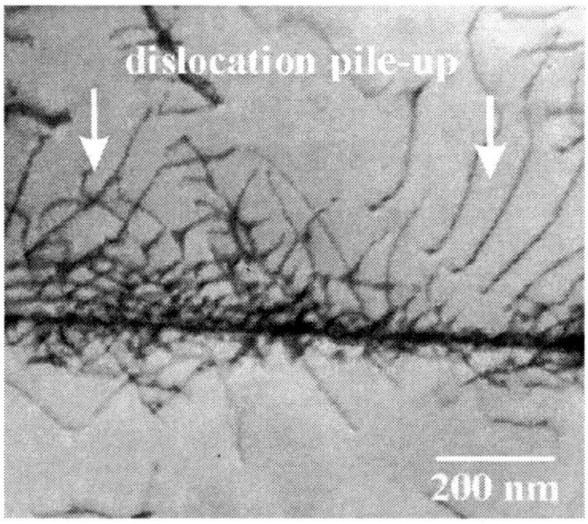

Fig. 6.23 TEM image of dislocation pile-up at a grain boundary leading to a quasi brittle mechanical response of the material.

If cyclic loading is applied to a material rather than static loading, the fatigue loads are less than the corresponding static loads. This leads to an important consequence: the material **fails at a lower stress level** than at a (single) tensile test. Note that cracks typically initiate at high stress concentration sites like corners or notches, which leave distinct markings at the fracture surface, due to plastic deformation caused by crack propagation.

How much a crack propagates depends on the frequency and amplitude of the loading cycle, but not on the load value itself.

Apart from the load and number of cycles, *other parameters* like atmosphere or temperature and impurity content affect the fatigue test.

The crack can propagate upon cyclic loading, which means there is a change in crack length with the change of loading condition, i.e., with the change of the number of cycles. This represents a **dynamic condition** (meaning the experimental conditions are not fixed) and can be expressed as a **power-law dependence** (with A and m being constants):

$$\frac{da}{dN} = \frac{crack\ length\ change}{change\ in\ number\ of\ cycles} = A \cdot \Delta K^m$$

Substitution of $\Delta K = Y\sigma\sqrt{\pi a}$ and subsequent integration gives the number of cycles to failure, N_F:

$$N_F = \frac{a_f^{-(m/2)+1} - a_0^{-(m/2)+1}}{A\sigma^m \pi^{m/2} Y^m \left[-(m/2)+1\right]} \quad (6.16)$$

Note that there is no unique theory about fatigue failure.

6.7 Creep

The tensile test did not tell the full story about materials used under cyclic loading conditions. Similarly, the tensile test does not allow the prediction of materials behavior at *high temperature*.

Creep is the **time-dependent strain**, which is especially prevalent at **high temperature**. **Creep** can be *defined* as plastic (permanent) deformation occurring at high temperature under constant load over a long time period. The corresponding creep rate is given by:

$$\frac{d\varepsilon}{dt} = creep\ rate \quad (6.17)$$

The elastic strain induced in a typical metal bar loaded below its yield strength at room temperature can be calculated from Hooke's law. This strain will generally *not change with time* under a fixed load. However, when repeating this experiment at high temperature (T being greater than one-half or one-third the melting temperature, T_m) produces a drastically different result.

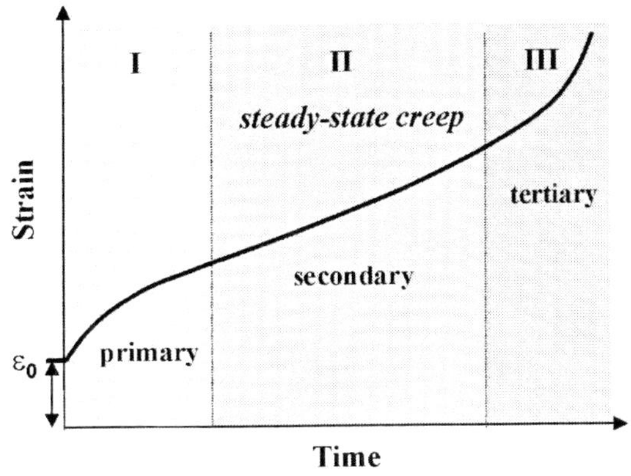

Fig. 6.24 Typical creep curve obtained by high-temperature deformation of a material under constant load over a long time period (about 1/2 or 1/3 of the melting temperature, T_m).

Commonly a **gradual increase** in **strain with time** is observed after the initial elastic loading. There are three distinct regions in a creep curve (primary, secondary and tertiary), as shown in Figure 6.24.
In each of these regions different creep mechanisms are active.

- In the **primary** creep region, the creep rate *decreases* with time and the material is *strain hardening.*
- In the **secondary** creep region, which is the longest and also called **steady-state creep**, the *dislocation mobility* is greatly enhanced, which balances the initial effect of strain hardening.
- In the **tertiary** creep region, *void formation* at grain boundaries occurs just prior to ductile fracture (metals). ***Note*** that all these mechanisms occur at elevated temperatures below the yield stress of the metal.

Note that creep deformation usually occurs at about 1/3 or 1/2 of the melting point of the material. However, materials such as ice, rock salt or even glass can creep at *room temperature*.

Since creep is a **thermally activated** process, one can apply the Arrhenius approach and determine the activation energy of creep deformation (Figures 6.25 and 6.26).

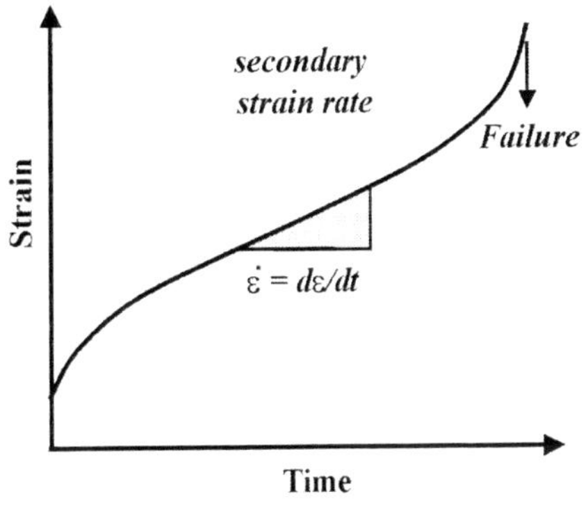

Fig. 6.25 Schematic showing how to determine the strain rate of the steady-state creep regime.

In Figure 6.26, the corresponding Arrhenius plot of creep rate, $\ln\dot{\varepsilon}$, versus the inverse temperature, $1/T$, is shown. Such a plot allows the determination of the *activation energy* of the respective creep mechanism. However, it should be emphasized that when different creep mechanisms are active simultaneously, the slowest process will be rate determining (similar to diffusion, where volume diffusion is the slowest process as compared to grain boundary and surface diffusion).

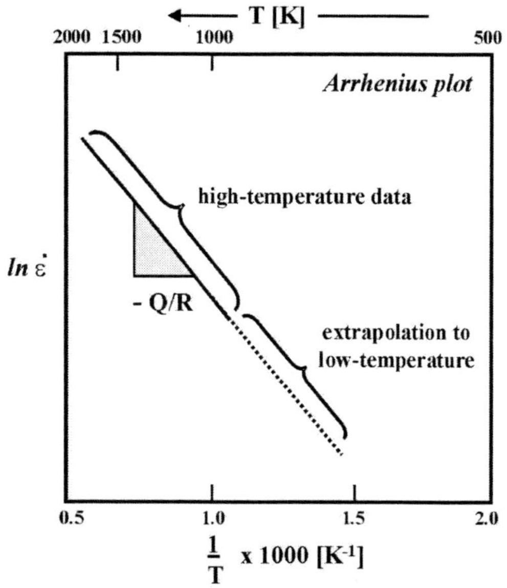

Fig. 6.26 Arrhenius plot of *ln* creep rate (secondary regime) versus 1/T. The slope of the line equals –Q/R.

6.8 Recovery and Recrystallization

Apart from high-temperature creep deformation, there are also other mechanisms active at high temperatures: **recovery** and **recrystallization**. These mechanisms are very important for *processing of metals*. After cold working a metal (rolling a thin metal sheet or pressing a metal part), one needs to soften the material by reheating in order to *release residual stress*. The technical term of this reheating process is **annealing**.
During **recovery**, some of the internally stored *strain energy* is relieved by *dislocation motion*, due to enhanced atomic diffusion at elevated temperature. The total number of dislocations (dislocation density) is reduced during recovery, and dislocation configurations with low strain energies are established.

Even when recovery is nearly complete, there will still be grains in the microstructure that have relatively high strain energies. **Recrystallization** is defined as the formation of a *new set of strain-free* and equiaxed grains. The *driving force* for the process of new-grain formation is the difference in *internal energy* between the strained and unstrained matrix grains. Initially, small nuclei are formed that grow until they completely replace the parent grain; a process that involves short-range diffusion.

The recrystallization mechanism is affected by various parameters:

- Amount of prior deformation (dislocation density)
- Temperature
- Time
- Initial grain size
- Composition.

The recrystallization process is not precisely defined, but can be quantified by:

$$t = C \exp(+Q/RT) \qquad (6.18)$$

with t being the time to complete recrystallization and C being a pre-exponential constant. ***Note*** that here Q is positive since the overall energy of the system is lowered and that this equation can only be used when calculating for a given percentage of recrystallization.

Example 6.2

If it takes 80 hrs to completely recrystallize an Al sheet at 250 °C and 6 hrs at 300 °C. Calculate the activation energy in kJ (kilojoules) per mol for this recrystallization process.

$$t_1 = C \cdot \exp(Q/RT_1)$$
$$t_2 = C \cdot \exp(Q/RT_2)$$

with $t_1 = 80$ h and $t_2 = 6$ h

Note that Q is *positive* here, since the overall energy of the system is lowered.

$T_1 = 250\ °C + 273 = 523$ K
$T_2 = 300\ °C + 273 = 573$ K

$$\frac{t_1}{t_2} = \frac{\exp(Q/RT_1)}{\exp(Q/RT_2)}$$

$$\frac{80h}{6h} = \exp\left\{\frac{Q}{R}\left(\frac{1}{523} - \frac{1}{573}\right)\right\}$$

$$\ln(13.33) = \frac{Q}{R}(0.001912 - 0.001745)$$
$$= \frac{Q}{R}(0.000167)$$
$$R = 8.314 \cdot 10^{-5}\ J/(mol/K)$$

$$Q = \frac{(2.59)(8.314 \cdot 10^{-5}\ J/(mol/K))}{0.000167}$$

$$= 128\ 900\ J/mol = 129\ kJ/mol.$$

Note that work hardening strongly increases the dislocation density and results in an increase in tensile strength, which, however, will finally cause brittle fracture of the component. Moreover, annealing the material after work hardening reduces the stress level and simultaneously affects strength, hardness (both are lowered) and ductility (which increases; Figure 6.27). In addition, a secondary heat treatment results in the formation of nuclei (small new grains), which continuously increase in size with time.

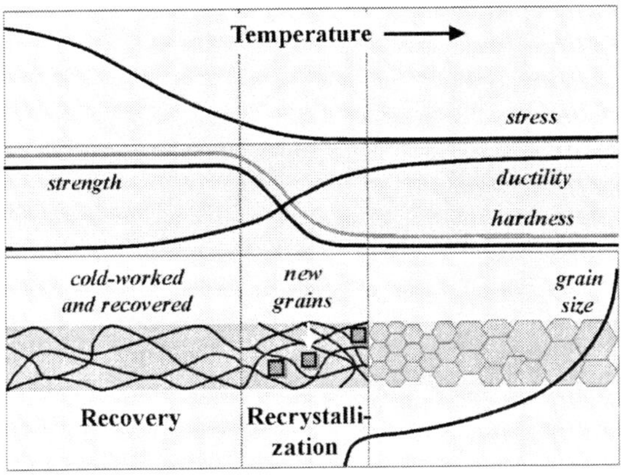

Fig. 6.27 Schematic illustration of recovery and recrystallization. *Note* the inverse correlation between ductility and hardness/strength of a material.

7 Engineering Alloys

Metals play a major role in engineering design, especially as structural elements. More than 90 wt% of the materials used for engineering are iron-based or ferrous alloys. For example, 98 % of the machinery made by Caterpillar is iron-based. These ferrous materials include steels, which contain 0.05–2.0 wt% C, and cast irons, with 2.0–4.5 wt% C. Most steels involve only a small amount of alloy additions to maintain moderate costs. Note that steel is the most *recycled* material, as it can be used over and over again (tens of millions of tons per year).

An **alloy**, unlike a composite with an interpenetrating network of two different phases, is composed of just one phase either as a **solid solution** or in some cases as a very fine dispersion of small **precipitates** of a separate phase within the host metal (examples are $AuCu_3$ and $Fe+Fe_3C$).

There are *four major classes* of ferrous alloys:

- plain carbon (ferritic; C < 2 wt%)
- low alloy steel (martensite and other mechanisms)
- high alloy (extensive alloy addition)
- cast iron (C > 2 wt%).

Processing is the key element in materials engineering (changing grain shape, chemical composition and microstructure). Special care in alloy selection and processing conditions can result, for example, in high-strength low-alloy (HSLA) steels. A composition of 5 wt% of non-carbon addition is typically used to define the *arbitrary boundary* between *low-alloy* and *high-alloy steels*. Table 7.1 gives the 6 generic ferrous alloys (all steels).

Cast irons exhibit a wide range of behavior, depending on their composition and their processing history (see also section 7.1.2).

Tab. 7.1 Composition and common application of the six generic ferrous alloys.

Metal	Composition	Uses
Low-C steel *(mild)*	Fe+0.04-0.3C +~0.8 Mn	*low* stress uses: constructional steel, welding application
Medium C-steel	Fe+0.3-0.7C +~0.8 Mn	*medium* stress uses: machinery parts, nuts, bolts, shafts, gears
High-C steel	Fe+0.7-1.7C +~0.8 Mn	*high* stress uses: springs, dies, cutting tools
Low-alloy steel	Fe+0.2C +0.8Mn1Cr 2Ni	*high* stress uses: pressure vessels, aircraft parts
High-alloy steel	Fe+0.1C +0.5Mn18Cr 8Ni	*high temperature* or *anti corrosion* uses; chemical or steam plants
Cast iron	Fe+2-5C +0.8Mn2Si	*low stress* uses: cylinder blocks, drain pipes

Note that the first 3 categories in Table 7.1 above are plain carbon steels, while high alloy steels are stainless steels.

The addition of chromium produces **stainless steels** with high corrosion resistance. Additions such as tungsten lead to high hardness alloys commonly used as tool steels (compare section 7.1.1). **Superalloys** include many stainless steels that combine corrosion resistance with high strength even at elevated temperatures.

One typically distinguishes between two types of **engineering alloys**:

- Ferrous alloys
- Non-ferrous alloys (non-Fe-based).

7.1 Ferrous Alloys

The majority of ferrous alloys are carbon and low-alloy steels. The main reason for this is that such alloys are moderately priced due to the absence of large quantities of expensive alloying elements and that they are sufficiently ductile to be readily formed into complex shapes by rolling, bending, stamping, forging, extrusion or drawing. Hence, ferrous alloys find applications like ball bearings and metal sheets for automobile bodies.

A convenient designation system for these alloys is the AISI-SAE system (American Iron and Steel Institute – Society of Automotive Engineers); for example 52150. The first two numbers give a code designating the type of alloy additions and the last two or three numbers give the average carbon content in hundredths of a weight percent.

As an example, for low alloy steel or plain carbon steel a 4-digit number is used: XXXX. A plain-carbon steel (10) with 0.40 wt% C is a **1040** steel, whereas a steel with 1.45 wt% Cr and 1.50 wt% C is a **52150** steel. In the latter example, the first two number indicate 52=Cr addition, while the last three numbers again refer to the carbon content: 150=1.5 wt% carbon.

It is important to realize that the addition of carbon to iron *changes* the respective **microstructure** and the corresponding **properties** of the ferrous alloy. Therefore, it essential for an engineer to understand the correlation between microstructure and properties, which in turn is closely related to the Fe-C phase diagram shown in Figure 7.1.

Note that depending on the volume fraction of carbon added to the system, different phases and their respective distribution within the matrix evolve, which leads to markedly different mechanical properties.

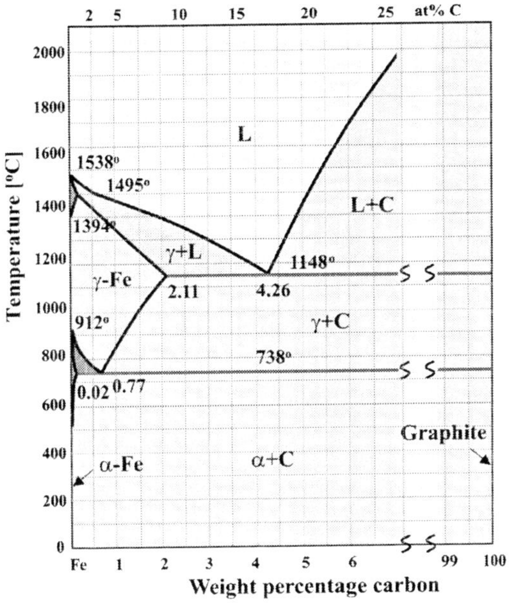

Fig. 7.1 Fe–C phase diagram (eutectoid).

The boundary between *irons* and *steels* is at **2.0 wt% C**, which roughly corresponds to the point of the **carbon solubility limit** in the **austenite**, γ-Fe, phase field (2.11 wt% C; Figure 7.2).

The **eutectoid reaction** of the Fe-C system:

$$\gamma - Fe \Rightarrow \alpha - Fe + Fe_3C \qquad (7.1)$$

produces a **pearlite** microstructure (Figure 7.3), which is composed of alternating layers of α-Fe and Fe_3C and lies close to the composition of a plain-carbon steel (1080 steel).

Note that the cementite phase, **Fe_3C**, and not carbon represents the second component in this system (Figure 7.2). The composition axis is customarily given in wt% C. Although graphite is a more stable precipitate than Fe_3C, the rate of graphite precipitation is much slower than that of cementite. Such a behavior is described by the **Ostwald rule** [29], which says that metastable phases are formed prior to the formation of the thermodynamically stable phase. As a result, many steels or cast irons contain the metastable Fe_3C phase.

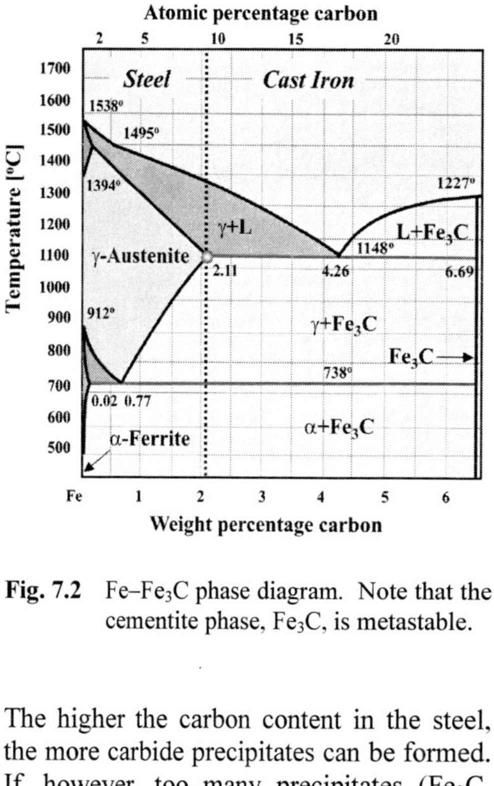

Fig. 7.2 Fe–Fe_3C phase diagram. Note that the cementite phase, Fe_3C, is metastable.

The higher the carbon content in the steel, the more carbide precipitates can be formed. If, however, too many precipitates (Fe_3C, Cr_7C_3) are present, the material becomes less ductile and may fail in a rather brittle manner.

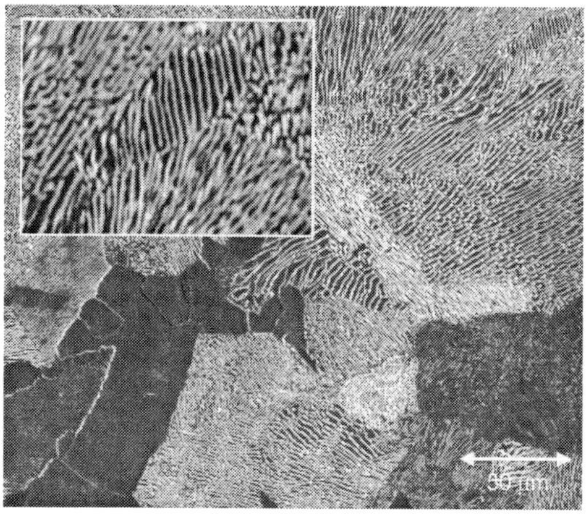

Fig. 7.3 Typical pearlite microstructure with alternating layers of α-Fe (dark gray) and Fe_3C (bright contrast).

Moreover, such carbide precipitates can grow and coalesce when the material is used at elevated temperature and can cause catastrophic fracture (large precipitates act as large intrinsic flaws).

As just noted, the Fe-graphite system is actually more stable, but due to its rather **slow kinetics** it is less common. However, very slow cooling rates can produce microstructures with precipitates of graphite. A more practical way to yield graphite formation is to add a small fraction of a third component such as silicon (about 2-3 wt%), which results in the formation of gray cast iron (see also section 7.1.2).

Note that the ability to change the *intrinsic microstructure* of a material by changing *processing parameters* bears great potential for engineering required materials properties for specific applications such as high hardness and/or high strength.

[29] **Wilhelm Ostwald** (1853-1932), Latvian chemist who received the Nobel prize for chemistry in 1909.

7.1.1 Stainless Steel

There is a wide variety of stainless steels with different composition, microstructure and resulting properties. For example, the incorporation of carbon into the *crystal lattice* (which is different from carbide precipitation) changes the (i) **lattice parameters** and (ii) the material **properties** (like hardness), as shown in Figure 7.4.

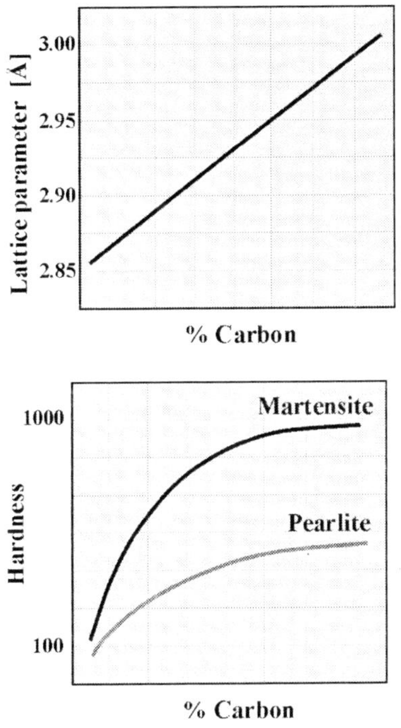

Fig. 7.4 Variation of lattice parameter, a_0, and hardness, H, with increasing carbon content (Fe-C system).

We will distinguish between:

- **Ferritic**; Fe-Cr binary alloys with 12-30 wt% Cr addition. Their structure is predominantly ferritic (bcc α-Fe). **Cr** also has the **bcc** structure and stabilizes ferritic steel. High-temperature applications range from kitchen hoods to combustion chambers.

- **Austenitic**; Fe-Cr-Ni ternary alloys with 16-25% Cr and 7-20% Ni addition. The presence of **Ni** stabilizes the γ-iron (austenite) structure down to room temperature. Austenitic steels reveal high formability due to their **fcc**-structure. Applications are chemical and food processing equipment, pressure vessels, and tank cars for aggressive chemicals.

- **Pearlitic**; used for rail, patented wire (cold drawn), and cables (very strong).

- **Precipitation hardened** stainless steels represent another class (compare also Figure 7.6).

- **Martensitic**; Fe-Cr binary alloys with 12-17 wt% Cr and a C-content ranging between 0.15-1.0 % C, which stabilize the **martensite** phase. Good strength and hardness *at sacrifice* of corrosion resistance. Applications are machine parts, pump shafts, valves, cutlery, bearings, surgical tools, or ball bearings.

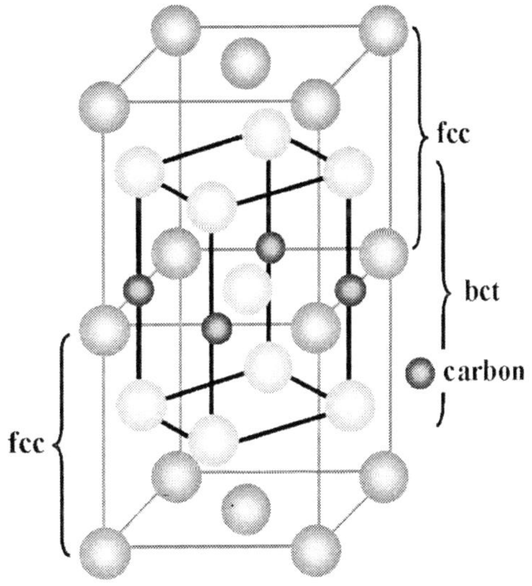

Fig. 7.5 Comparison between the fcc-austenite and bct-martensite crystal structure.

Note that *martensite* is a *quenched* austenite structure with a **body-centered tetragonal** (bct) Bravais lattice (Figure 7.5).

7.1.2 Cast Irons

Cast irons contain 2-4 % C and commonly ~1% Si to stabilize (or kinetically enhance) the formation of the graphite phase. One distinguishes between the following four types of cast irons:

- **White** cast iron shows a white fracture surface, since carbon forms **Fe_3C** precipitates instead of graphite. The formation of Fe_3C results in good **wear resistance**.

- **Gray** cast iron is processed with the addition of a small amount of **Si** (2-3 wt%), which supports the formation of **graphite flakes**. Gray cast irons show good machining and vibration dampening ability and are used as cylinder blocks, clutch plates, and heavy gear boxes.

- **Ductile** cast iron contains a high volume fraction of ferrite and is therefore rather ductile. By adding a small amount of Mg (0.05 wt%) **spherical graphite nodules** are formed. Ductile cast irons are commonly formed via pressure casting and used as valve and pump bodies, crankshafts, and gears.

- **Malleable** cast iron is initially cast as white iron and then *heat treated* to form **graphite nodules** as rather small precipitates. Malleable cast irons show good uniformity and are used as connecting rods or universal joints.

7.2 Non-Ferrous Alloys

There are *two major types* of non-ferrous alloys designated by XXX.X and XXXX. The **dot** indicates that it is a **cast alloy** (pouring a liquid metal into a sand bed). This type of processing causes many defects like voids, porosity, surface flaws etc., which can be good or bad depending on the specific application (pores can act as stress concentrators or can host internal lubricants; *self-lubricating systems*).

The **4 digit** number (no dot, XXXX) indicates it is a **wrought alloy**. The first digit represents the composition of the alloy (here Al-alloy):

1 = $\geq$ 99.00 Al	**6** = Mg+Si
2 = Cu	**7** = Zn
3 = Mn	**8** = other elements
4 = Si	**9** = unused series
5 = Mg	

Moreover, the **tempering conditions** are indicated by additional letters:

F = as-fabricated,
O = fully annealed,
H = strain hardened,
T = heat treated.

As an example, the Al-alloy **2024-T6** is commonly used for aircraft parts and larger structures for aircrafts. The **20** indicates that the alloy group is Cu, **24** gives further identification of the specific alloy composition: 4.4% Cu, 1.5% Mg, 0.6% Mn, and **T6** stands for a six hour heat treatment.

7.2.1 Al-Alloys

Al-alloys are best known for their **low density,** $\delta = 2.70\, g/cm^3$, and good **corrosion resistance**. The importance of Al within the family of metals has increased due to its low density. For example, the total mass of a new American car dropped by nearly 20 % between 1980 and 2000 due to an increased use of Al (in addition to polymers). Al-alloys are commonly used in aircraft, automobiles, beverage cans, and lightweight constructions (light poles, windows, sidings). Apart from its low density, aluminum shows high conductivity (thermal and electrical) and is therefore used as a heat exchanger or in electrical applications.

Note that initially, the steel industry did not expect a strong competition from the aluminum industry and hence did not pursue research in this area. Nowadays, however, steel industry is trying to get back into the Al-can business.

Since aluminum has the **fcc** structure, it is *inherently ductile* (easy to form and mold due to the 12 slip systems that can be activated). If we consider a given metal such as Al, it is very important for an engineer to know how to *adjust mechanical properties* to *specific application needs*. For example, what has to be done in order to ***strengthen a metal***? One has to *reduce or limit dislocation motion* by either

- **work hardening** *or*
- **precipitation hardening.**

Work hardening is a process where the metal is mechanically deformed, commonly at rather low temperature (*cold working*). Due to the deformation process, a high density of dislocations is generated, which leads to an increase in hardness and strength of the metal. Note that an annealed metal can have a dislocation density as low as $10^{10}\, m^{-2}$, while a cold worked metal typically reveals dislocation densities of 10^{18}-$10^{20}\, m^{-2}$. Since the number of dislocations in the deformed metal is so high, dislocation motion is highly limited and dislocation interaction occurs. In particular, the reduced ability for dislocations to move increases both hardness and strength.

Another way to *change* the *yield stress* of an Al-alloy is by **precipitation hardening.** This process can be rationalized as follows: the small particles formed within the matrix act as *obstacles* to *dislocation motion*, which, similarly to the work hardening mechanism, increases hardness and strength of the metal. During dislocation motion, the dislocations start to bend around the obstacle/precipitate and are locally pinned (Figures 7.6 and 7.7).

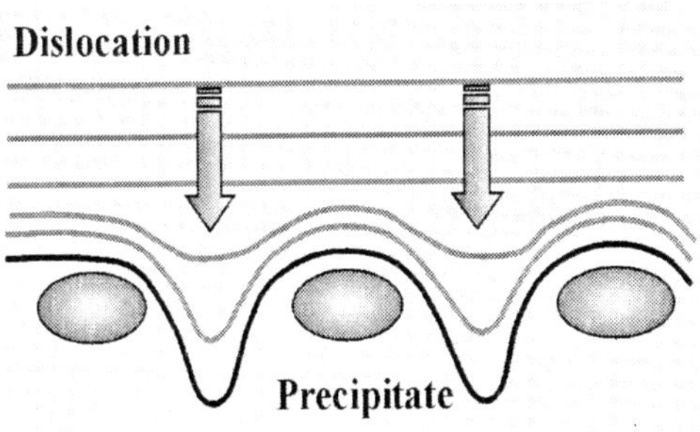

Fig. 7.6 Schematic illustrating the mechanism of precipitation hardening due to the formation of small precipitates within the host material, lowering dislocation mobility.

Note that a similar mechanism applies for **solid solution** alloying, where the incorporated atoms/ions can also act as obstacles since they create a wider strain field, which again lowers dislocation mobility and increases hardness and strength. Hence, *alloys* are commonly *harder* than the corresponding pure metals.

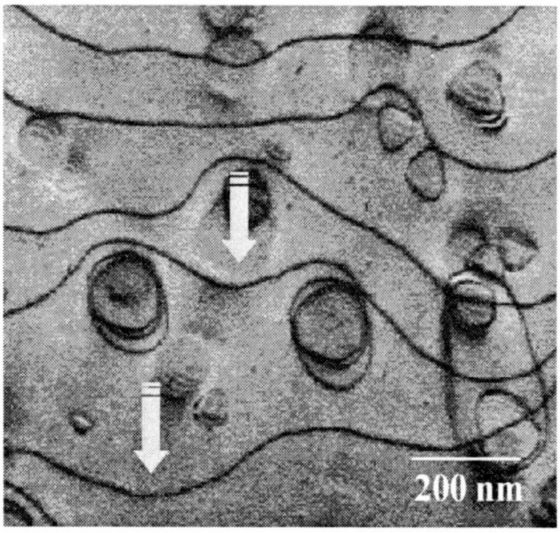

Fig. 7.7 TEM micograph of disclocations (dark lines) bending around local obstacles (precipitates).

Note that complex structures tend to be stronger but less ductile due to their limited number of slip systems. Moreover, breaking covalent and/or ionic bonds is more difficult than breaking metallic bonds. As a result, ceramics are hard and strong but brittle.

In order to explain how to prepare *precipitation hardened aluminum*, we will look at a generic Al-X phase diagram (Figure 7.8).

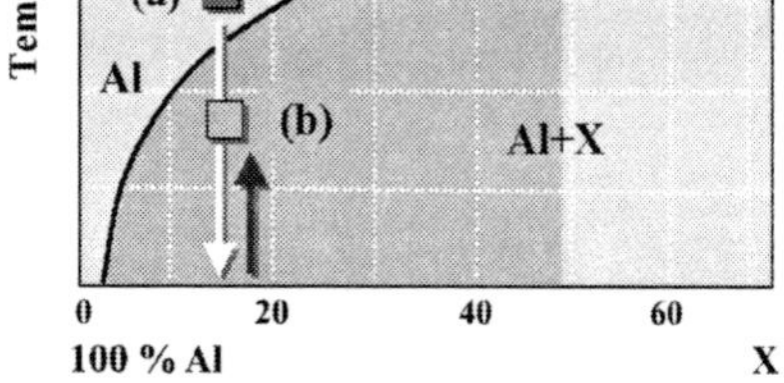

Fig. 7.8 Generic Al-X phase diagram used to explain how to process a precipitation hardened Al-alloy.

The processing steps involved to prepare a precipitation hardened Al-alloy are:

1. **solutionizing** at point (a)
2. **quenching** (cooling down quickly) to room temperature
3. **aging**; bring to a certain temperature and hold at (b).

A specific recipe for an Al-Cu alloy (Al-4 wt% Cu; 2024-T6) is given below:

1. **solutionizing** at 515 °C
2. **quenching** in H_2O and cooling down to room temperature
3. **aging** at 130–190 °C for 6 hours.

The corresponding properties of the 2024-T6 Al-alloy have markedly changed with aging and/or annealing:

as-processed: σ_{UTS} = **220** MPa
σ_Y = **97** MPa

aged: σ_{UTS} = **442** MPa
σ_Y = **345** MPa.

Note that since the aging temperature is rather low, there is a potential danger that the material overages when being used at higher temperatures (growth and coalescence of precipitates forming large intrinsic flaws = fracture origin).

Below are examples of important Al-alloys and their corresponding applications and properties:

- **6061-T6** = 1.0% Mg, 0.6% Si, 0.27% Cu, 0.20% Cr; truck and marine structures, σ_{UTS} = **290** MPa
- **356.0** = cast Al, 7% Si, 0.3% Mg; transmission cases, wheels, σ_{UTS} = **210** MPa

- **355.0** = 5% Si, 0.5% Mg, 1.2% Cu; Si increases fluidity during casting and increases strength; pump housings, crank cases, σ_{UTS} = **220** MPa

- **3004-H38** = 1.2% Mn and 1.0% Mg; beverage cans, σ_{UTS} = **220** MPa.

Note that precipitates often form at the *surface* of the material. One example is frost (small ice crystals) at the surface of grass on a cold morning (not seen in the air). When the temperature drops, the size of the ice crystallites is reduced.
Moreover, a *fine dispersion* of the precipitates has a stronger effect than the presence of only a few large precipitates within the matrix.

In conclusion, by alloying and processing (cooling rate, tempering) one can control microstructure evolution and thus materials properties.

7.2.2 Cu-Alloys

Copper alloys possess a number of superior properties such as excellent **electrical** and **thermal conductivity** as well as good **corrosion resistance** even in *marine environment*. Similar to aluminum, Cu-alloys are soft and show high ductility due to the fcc-structure (the density of Cu is $\delta = 8.96\,g/cm^3$).

There are basically two major Cu-alloys:

- **brasses**
- **bronzes**.

The combination of **Cu+Zn** is called **brass** (with 5-40 wt% Zn). Brass is very **machinable** and the more Zn the alloy contains, the higher the strength. Brass can also be **work hardened**, to further improve hardness and strength.

The combination of **Cu+Sn** is termed **bronze**. Bronze can easily be casted and has better overall properties than brass. Therefore, bronze is the main Cu-alloy used for artistic sculptures.

The **Cu-Zn** phase diagram (Figure 7.9) is rather complex, similar to the Al-Cu system; however, from a practical point of view it is rather easy to analyze.

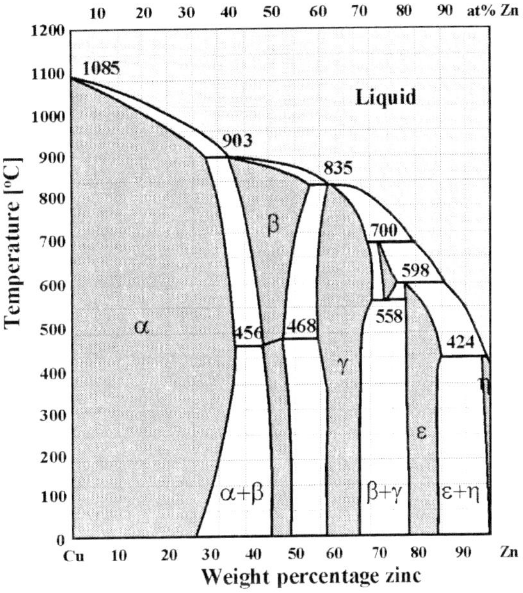

Fig. 7.9 Complex Cu-Zn phase diagram.

Common application of the various Cu-alloys given in Table 7.2 are:

- wire, water piping, roofing
- heat exchanger tubing (sometimes with 30% Ni)
- brasses: jewelry, ammunition casings, cartridges, architectural design (door knobs)

- bronzes (sculptures)
- major components in car radiators, music instruments and coins.

Tab. 7.2 Classification of various copper alloys.

Classification of Cu-Alloys	
	Wrought alloys
1xxxx	Cu*, high-Cu alloys**
2xxxx	Cu-Zn alloys (brasses)
3xxxx	Cu-Zn-Pb alloys (leaded brasses)
4xxxx	Cu-Zn-Sn alloys (tin brasses)
5xxxx	Cu-Sn alloys (phosphor bronzes)
6xxxx	Cu-Al alloys (Al bronzes), Cu-Si alloys (silicon bronzes)
7xxxx	Cu-Ni and Cu-Ni-Zn alloys
	Cast alloys
8xxxx	cast Cu, cast high-Cu alloys, cast brasses of various types, cast Mg-bronze , cast Cu-Zn-Si alloys.
9xxxx	Cast Cu-Sn alloys, Cu-Sn-Pb, Cu-Sn-Ni alloys, Cu-Al-Fe, Cu-Ni-Fe, Cu-Ni-Zn alloys

* Cu-content of 99.3% or higher
** less than 99.3% Cu, but more than 96%

7.2.3 Ti-Alloys

Ti-alloys have been widely used since World War II since they are relatively **light-weight** with a density of $\delta = 4.54\, g/cm^3$, but also rather **strong**. A 99.9% pure Ti metal has a yield strength of σ_y = **650** MPa.

On the other hand, Ti is **expensive** and difficult to process since it is very **reactive** with O, N, C, H, and Fe. However, once processed, its reactivity works to its advantage since a thin oxide coating on the surface (*passivation layer*) gives it good **corrosion resistance**.

Ti shares the **hcp** structure with Mg and Zn (Table 7.3), leading to a characteristically **low ductility** (brittleness). However, the *high-temperature bcc structure* can be stabilized down to room temperature by adding alloying elements such as vanadium, strongly increasing the ductility of Ti. Therefore, the Ti-alloy **Ti-6Al-4V** with an ultimate tensile strength of σ_{uts} ~**1000** MPa is the most important Ti alloy.

Tab. 7.3 Physical properties of various metals commonly used for engineering applications.

Physical Properties of Engineering Metals	Density [g/cm^3]	Melting Point [^{o}C]	Crystal Structure
Mg	1.74	651	hcp
Al	2.70	660	fcc
Ti	4.54	1675	hcp – bcc
Ni	8.90	1453	fcc
Zn	7.13	419	hcp
Fe	7.87	1535	bcc – fcc
Cu	8.96	1083	fcc

Applications of Ti-alloys are aerospace structures, sporting goods (bike frames, golf clubs), pumps and valves, as well as prosthetic joints since Ti is biologically inert (*biomaterials*).

7.2.4 Ni-Alloys

Ni-based alloys have a relatively **high density** of $\delta = 8.90\, g/cm^3$ and are rather expensive. However, Ni-alloys are **corrosion resistant** and show **high-temperature oxidation resistance**.

Ni-alloys are an important component of stainless steels and coins (the alloy used in coins is termed *monel*).
Note that the *penny* has a Zn-core and an electroplated Cu-surface. A *nickel* has no core/rim structure, but shows a homogeneous composition of Cu+25%Ni. A *dime* has a Cu-core and a Cu+25%Ni surface coating. The *dollar* is a Cu-Ni-Zn-Mn alloy, which allows control of both conductivity and color.

Ni-based **superalloys** (50-80% Ni, 15-20% Cr; 1-4% Al, 2-4% Ti) are typically composed of a **fcc** matrix of **γ-(Ni,Cr)** with small precipitates of $\mathbf{Ni_3Al}$ or $\mathbf{Ni_3Ti}$ within the matrix.

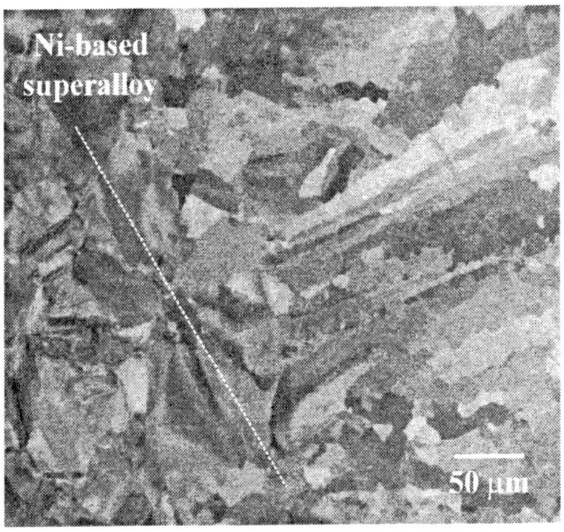

Fig. 7.10 Microstructure of the heat-affected zone of a welded region in a Ni-based superalloy.

Figure 7.10 shows the heat-affected zone of the weld in a Ni-based superalloy. The large columnar grains on the right grew during solidification while the deformed and twinned microstructure of the starting material is shown on the left.

7.2.5 Other Common Metals

Mg-alloys are **lightweight** due to the low density of Mg, $\delta = 1.74\, g/cm^3$. Since Mg, similar to Ti, has the **hcp** structure, it is inherently brittle. Applications of Mg-alloys are suitcase frames, pencil sharpeners, and computer casings. ***Note*** that there is only a limited supply of Mg; one potential source is the ocean.

Zn-alloys are rather heavy due to the high density of Zn, $\delta = 7.13\, g/cm^3$, but have a rather low melting temperature (Table 7.3). Zinc is heavily used as protective coatings on steels and compose the core of the *penny*.

Pb-metal is used as battery anodes, in bullets (high density of $\delta = 11.3\, g/cm^3$) and as radiation shielding. Note that Pb creeps at room temperature.

Refractory metals, W, Mo, Ta, Nb, have very high melting points and are therefore used as filaments in light bulbs (K-doped W), furnace heating elements (Mo) and superconductor material (Nb). ***Note*** that tungsten carbide, WC, embedded in a cobalt matrix is a metal-ceramic composite used for cutting tools.

Precious metals, Au, Ag, Pt, Ir, Os, Pd, are highly corrosion resistant and excellent electrical and thermal conductors. They are used as dental fillings (Au), in the electronics industry (Ag) or as catalysts (Pt; catalytic converter in cars). ***Note*** that Ag is used as a chemical for photo processing.

8 Oxidation and Corrosion

Oxidation and corrosion are the cause of failure of many engineering materials. Examples are rusting of cars and bridges, tarnishing of silver or copper, decomposition of concrete buildings or even the decay of teeth (cavity formation). In this chapter we will discuss the effect of oxidation and corrosion especially on metals. Although ceramics also suffer from corrosion and oxidation, which strongly depends on their composition, they tend to be more stable as compared to metals.

8.1 Oxidation

Oxidation is a chemical reaction between a metal and atmospheric oxygen. In general, metals as well as alloys form **stable oxide** compounds when exposed to air, in particular, at elevated temperature. A few notable exceptions such as the noble metals Os, Ir, Pt, Au are highly prized. ***Note*** that *pyrophoric* metals can react explosively when exposed to air (fine powders of Al, Mg, etc.). The stability of metal oxides is demonstrated by their high melting points compared to the pure metal. For example **Al-metal** melts at **660 °C** while its oxide **Al_2O_3** melts at **2054 °C**. Hence, the oxidation product is thermodynamically more stable than the corresponding metal.

When oxygen reacts with the metal surface, $Me+O_2 \rightarrow MeO$, MeO_2, Me_2O_3, Me_3O_4 etc., oxygen can either diffuse into the metal, metal ions can diffuse through the oxide scale to the surface, or counter-diffusion can take place, as illustrated in Figure 8.1. There are three kinds of oxide scales:

- porous
- adherent
- spalling.

The **porous** surface coating is caused by volume changes during oxidation; it is **not protective** and will result in a continuous oxidation of the metal surface (open porosity; Figure 8.2).

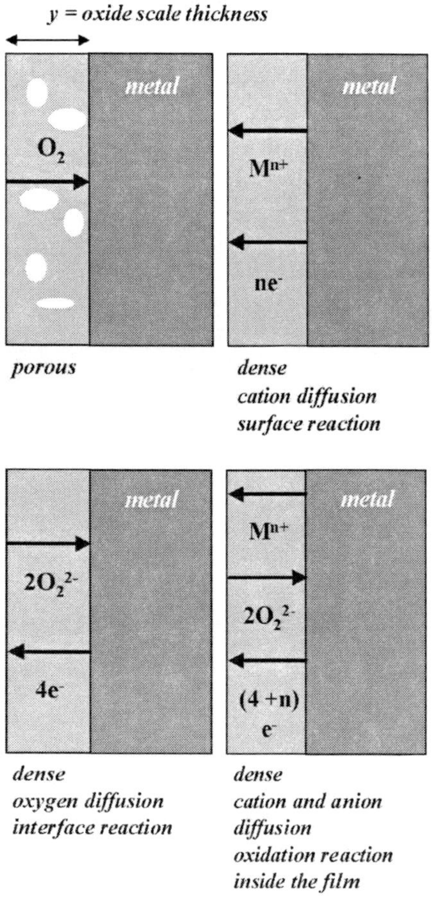

Fig. 8.1 Schematic showing the different diffusion paths possible when a metal surface is oxidized.

On the other hand, the **adherent** surface coating is **dense** and shows typically **good adhesion** to the metal surface. Such oxide scales are commonly **protective** and are therefore advantageous (like for example a thin coating of Al_2O_3 on Al or TiO_2 on Ti).

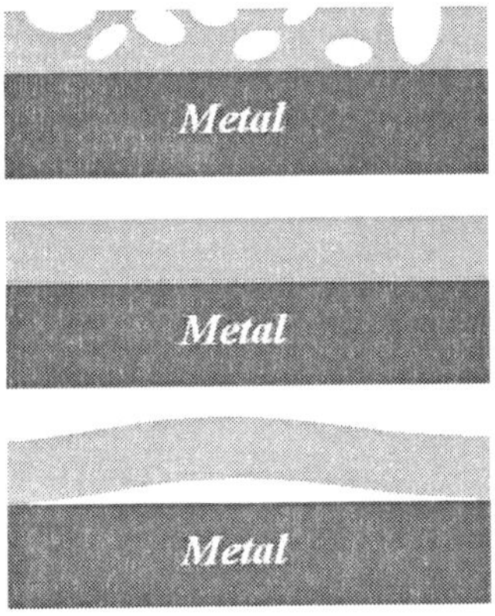

Fig. 8.2 Schematic of porous, adherent, and spalling oxide coatings.

Spalling of the surface coat is also rather detrimental for the metal because, although initially a dense coating is formed, spalling off (flaking) for example caused by compressive stresses, creates **new surface area** of the underlying metal which then oxidizes.

It would be very helpful for the engineer to be able to estimate what type of oxide scale is formed under a special application. The **density** of the oxide that is formed is one important parameter that determines the type of oxide scale. As a *rule of thumb* one can use the **Pilling-Bedworth ratio**, R, in order to determine the scale type:

$$R = \frac{Md}{xmD} \quad (8.1)$$

where R corresponds to the volume of oxide (M_xO_y) produced over the volume of metal consumed. Here, M equals the molecular weight and D the density of the oxide, m is the atomic weight of the metal and d is the metal density. The value of x in the denominator depends on the oxide formed, e.g., Fe_1O, x=1; Fe_2O_3, x=2; Fe_3O_4, x=3, etc.).

Depending on the value of R, porous, adherent, or a spalling oxide coat will form:

R < 1 porous
R = 1-2 adherent
R > 2 spalling.

Hence, for $1 < R < 2$ a "natural" protective oxide scale is formed like Al_2O_3 on Al or Fe-, Cr-oxides on stainless steel. ***Note*** that the simplified expression for R given in equation 8.1 becomes more complex, when for example a **Cu-Ni alloy** is oxidized. In this case, one has to consider the **free energy** of oxide formation and the **diffusion rate** of the different species involved.

Example 8.1

Calculate the Pilling-Bedworth ratio, *R*, for the formation of Tenorite, CuO, on a Cu-metal surface; use Tenorite density = 6.5 g/cm^3, Cu density = 8.96 g/cm^3 and the atomic weights of Cu and O = 63.55 and 16.0, respectively. What type of oxide scale is formed?

$$R = \frac{Md}{xmD} \qquad R = \frac{(63.55+16)\cdot 8.96\ gcm^{-3}}{1\cdot 63.55\cdot 6.5\ gcm^{-3}}$$

$$R = 1.726$$

Therefore, Tenorite forms an adherent and protective oxidation layer on the Cu surface.

8.1.1 Kinetics of Oxidation

The **kinetics** of an oxidation process describes how fast the reaction between oxygen and metal will occur and how fast the thickness of the oxide coat on the surface will increase (depending on time and temperature). For a more detailed under-standing of the growth kinetics of an oxide layer one has to consider thermodynamics, phase diagrams, free energy, and phase stability.

There are two general types of oxidation kinetics (see also Figures 8.3 and 8.4):

- **linear**: where oxygen is always available (porous coating):

$$y = k_l t \quad (8.2)$$

- **parabolic**: where oxygen diffusion controls scale growth (adherent layer):

$$y^2 = k_p t \quad or \quad y = k'_p t^{1/2} \quad (8.3)$$

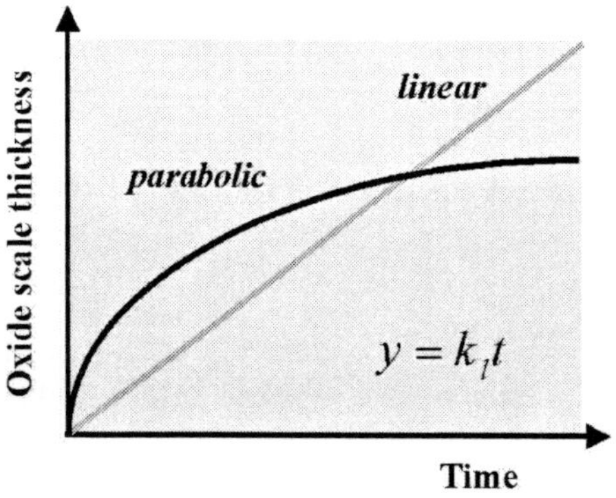

Fig. 8.3 Schematic of oxide scale thickness versus time of porous and adherent oxide coatings.

When the **diffusion** of oxygen through the oxide scale is not restricted (unlimited excess of oxygen at the interface), then the oxidation kinetics is **linear**.

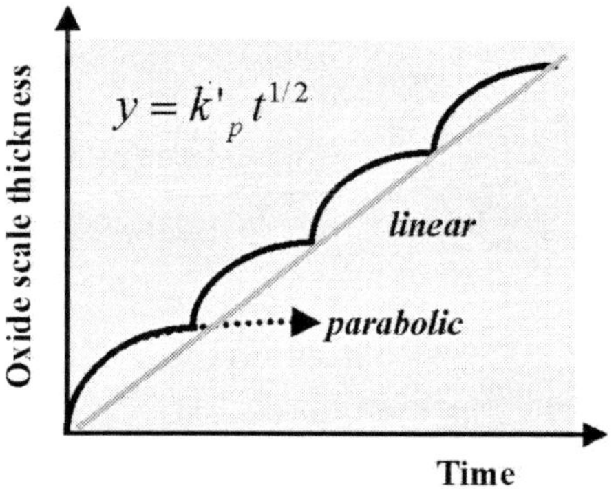

Fig. 8.4 Schematic of oxide scale thickness versus time of a spalling oxide coat.

However, when oxygen diffusion through the oxide coating is **limited**, oxidation first becomes slower and then stops completely, resulting in a **parabolic** behavior.

When **spalling** of the oxide scale occurs, the overall (long-term) oxidation kinetics of this process is linear, although on a shorter time scale a parabolic behavior is detectable (Figure 8.4).

Note that oxidation is an **electrochemical reaction**:

- **Oxidation**: $M \rightarrow M^{2+} + 2e^-$; with the metal being stripped of its electrons
- **Reduction**: $\frac{1}{2} O_2 + 2e^- \rightarrow O^{2-}$.

8.2 Corrosion

Corrosion is defined as the **deterioration** of a material due to **chemical attack** by its environment, which is commonly aqueous. Corrosion costs the United States infrastructure ≈50 bio$/yr; hence, is a very important engineering issue. The **main factors** affecting corrosion are:

- chemical composition
- concentration
- temperature.

Metals are very sensitive to corrosive attack, because they have free electrons that can move easily and thereby participate in **electro-chemical reactions** (oxidation). In contrast, glasses and polymers typically corrode by a direct chemical attack. Most ceramic materials are, however, chemically rather stable.

Tab. 8.1 Electromotive force series of noble (cathodic) and active (anodic) metals.

Electromotive Force Series

	$Me\text{-}Me^+$	Electrode potential	$Me\text{-}Me^+$	Electrode potential
noble or	$Au\text{-}Au^{3+}$	+1.498	$Co\text{-}Co^{2+}$	-0.277
cathodic	$Pt\text{-}Pt^{2+}$	+1.200	$Cd\text{-}Cd^{2+}$	-0.403
	$Pd\text{-}Pd^{2+}$	+0.987	$Fe\text{-}Fe^{2+}$	-0.440
	$Ag\text{-}Ag^+$	+0.799	$Cr\text{-}Cr^{3+}$	-0.744
	$2Hg\text{-}H_2^{2+}$	+0.788	$Zn\text{-}Zn^{2+}$	-0.763
	$Cu\text{-}Cu^{2+}$	+0.337	$Al\text{-}Al^{3+}$	-1.662
	½ H_2-H^+	**0.000**	$Mg\text{-}Mg^{2+}$	-2.363
active or	$Pb\text{-}Pb^{2+}$	-0.126	$Na\text{-}Na^+$	-2.714
anodic	$Sn\text{-}Sn^{2+}$	-0.136	$K\text{-}K^+$	-2.925
	$Ni\text{-}Ni^{2+}$	-0.250		

In order to understand corrosion in more detail, one needs to understand **oxidation-reduction reactions.** All these reactions have an **electrochemical potential**, E^0, which is commonly compared to the reference (standard) potential of hydrogen (see Table 8.1).

The oxidation-reduction reaction of iron that is in direct contact with hydrochloric acid, HCl, can be written as:

$$\mathbf{Zn^0 + 2HCl \implies ZnCl_2 + H_{2(g)}}$$

$$\mathbf{Zn^0 + 2H^+ \implies Zn^{2+} + H_{2(g)}}$$

The **oxidation** half of this reaction is:

$$\mathbf{Zn^0 \implies Zn^{2+} + 2e^-}$$

where the metal is stripped from its electrons (oxidation) and the corresponding **reduction** half of the reaction is:

$$\mathbf{2H^+ + 2e^- \implies H_{2(g)}}.$$

All such chemical reactions have an **electrode potential**, E^0, since electrons flow during the reaction (see also Table 8.1). The standard or reference potential is the above mentioned "reduction half" of the Zn+HCl-reaction, however, going the opposite direction:

$$\mathbf{½\, H_{2(g)} \implies H^+_{(g)} + e^-}$$

with an electrochemical potential of $E^0 =$ 0.000 volts.

An electrochemical corrosion reaction can be described as the formation of a **galvanic cell** between different materials (see Figures 8.5 and 8.6). ***Note*** that rusting of steel is also a galvanic corrosion, but does not form a galvanic cell.

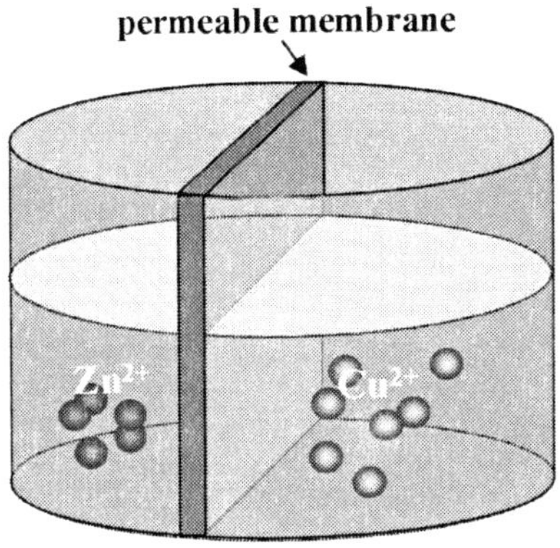

Fig. 8.5 Setup of a galvanic cell; *first step* (for details see text).

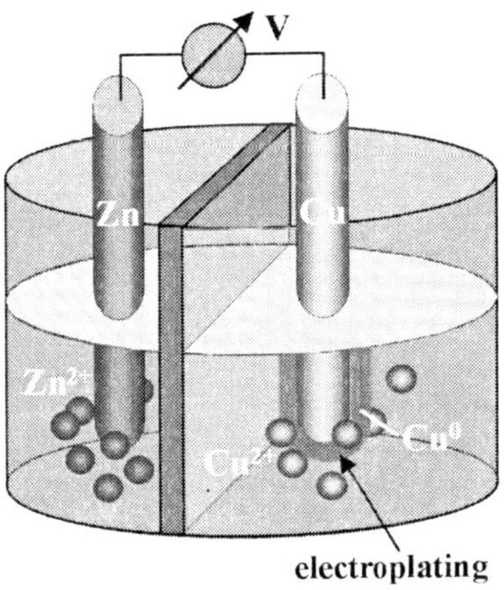

Fig. 8.6 Setup of a galvanic cell; *second step* (for details see text).

1st step: A galvanic cell is a beaker with a **permeable** (porous) **membrane** that allows ions to pass through it (to compensate for the variation in ion concentration). On either side of the membrane is a solution of metal ions, in this case of Zn^{2+}-ions (concentration of 1 molar $ZnSO_4$) and Cu^{2+}-ions (concentration of 1 molar $CuSO_4$).

2nd step: Now a pure Zn electrode is inserted on the left hand side and a pure Cu electrode is inserted on the right side of the galvanic cell. The two electrodes are then connected and a voltmeter is installed. This is the typical setup of a **battery**. When the two electrodes are connected, electron will flow from the side of the less noble electrode to that of the more noble metal, where they combine with the cathodic metal ions (compare electromotive force series in Table 8.1).

In the given scenario, the more noble metal is copper and the less noble metal is zinc. ***Note*** that the more **noble** metal, which will be reduced, acts as the **cathode**, while the **less noble** (active) metal is oxidized during the electrochemical reaction and acts as the **anode**; *"redcat" = reduced at cathode.*

Using Table 8.1 we find that zinc acts as the anode that is oxidized and that copper acts as the cathode, where the reduction reaction takes place:

$$\mathbf{Zn^0 ==> Zn^{2+} + 2e^-;\quad E^0 = -0.763\ V}$$

$$\mathbf{Cu^{2+} + 2e^- ==> Cu^0;\quad E^0 = -0.337\ V}$$

$$\mathbf{Zn^0 + Cu^{2+} => Zn^{2+} + Cu;\quad E^0_{cell} = -1.100\ V}$$

Note that here the electrochemical reaction of Cu ($Cu^0 => Cu^{2+} + 2e^-$) is reversed (cathodic reaction) leading to a negative sign.

In Shackelford's book, the reaction shown above is written differently:

$$\mathbf{Cu^0 => Cu^{2+} + 2e^-;\quad E^0 = +0.337\ V\quad cathode}$$

$$\mathbf{Zn^0 => Zn^{2+} + 2e^-;\quad E^0 = -0.763\ V\quad anode}$$

$$\mathbf{E^0_{cath} - E^0_{anode} = E^0_{cell};\quad E^0_{cell} = +1.100\ V}$$

Here, the **electromotive force** (Emf) of the galvanic cell is positive. The electrochemical reaction described above, results in **electroplating** of Cu onto the cathode, while Zn goes into solution. However, with time the Cu^{2+}-ions in solution run out and the voltage drops to zero, terminating the electroplating process.

Note that corrosion can also occur when dipping only **one electrode** into a solution. In this case, the formation of a local anode and cathode drives the reaction (e.g. dipping a Zn electrode into a HCl solution; see Figure 8.7). Such a reaction is still called **galvanic corrosion**. The corresponding anodic and cathodic reactions are given by:

- **anodic** reaction (oxidation)
  $Zn ==> Zn^{2+} + 2e^-$
- **cathodic** reaction (reduction)
  $2H^+ + 2e^- ==> H_2$.

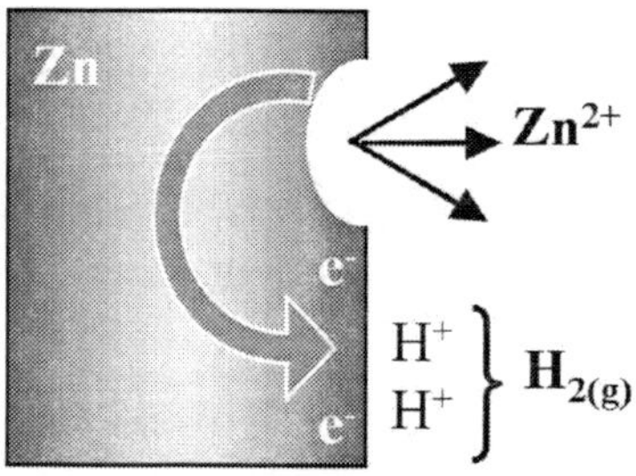

Fig. 8.7 Galvanic corrosion of a single electrode (zinc); for example locally in direct contact with HCl.

Another example of a single electrode galvanic corrosion is **rusting of iron**. When the iron is, for example, immersed in water, following reaction occurs:

- **anodic** reaction (oxidation)
  $2Fe \Rightarrow 2Fe^{2+} + 4e^-$, -0.440 V
- **cathodic** reaction (reduction)
  $O_2 + 2H_2O + 4e^- \Rightarrow 4OH^-$, -0.401 V

$$2Fe + 2H_2O + O_2 \Rightarrow 2Fe^{2+} + 4OH^- \Rightarrow 2Fe(OH)_2$$

In this case, OH^- reacts with Fe^{2+} (by-products formed in the above anodic and cathodic reactions), and forms iron hydroxide. $\mathbf{Fe(OH)_2}$ further oxidizes (reaction with OH^-) to $\mathbf{Fe(OH)_3}$ to form **rust** (the red-brown color of rust originates from $Fe(OH)_3$ precipitation; Figure 8.8)).

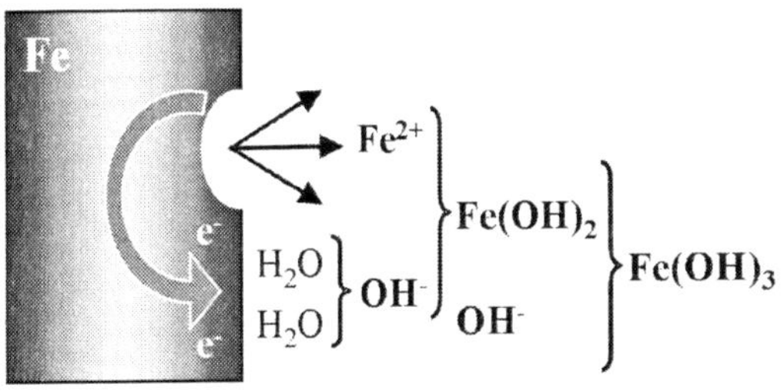

Fig. 8.8 Galvanic corrosion of a single electrode (iron), which is in contact with water.

8.2.1 Types of Corrosion

In principle, one can distinguish *seven* important types of different corrosion processes:

- **Uniform** *or* **general attack corrosion**
  Here, the electrochemical reaction proceeds uniformly across the entire surface of the component.

- **Galvanic** *or* **two-metal corrosion**
  If two metals are placed in direct contact, electrochemical reaction may occur depending on the difference in their electrochemical potentials (electro-motive force (Emf)). *Note* that such a reaction can also be used to protect steel by coating it with zinc (by hot dipping).

  Emf - Fe = -0.440 V
  Emf - Zn = -0.763 V

  In this system, **Fe** acts **cathodic** with respect to the Zn. Hence Zn will corrode first, thereby protecting steel. When **Sn** is used instead of Zn, tin acts as the more noble metal and iron will corrode.

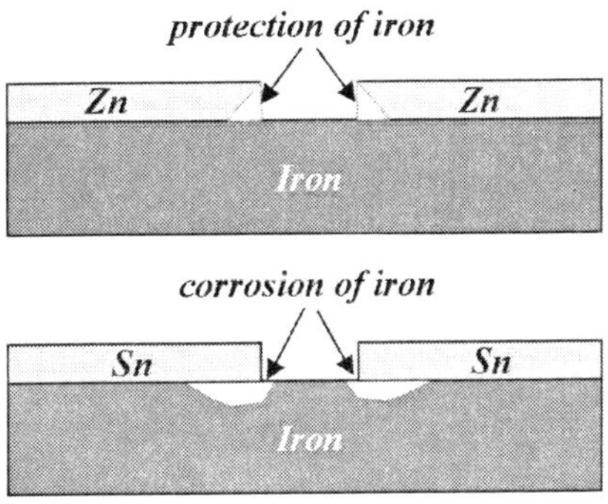

Fig. 8.9 Galvanic corrosion of two metals in direct contact. Zn protects Fe while Sn acts cathodic when O_2 is present.

Note that in *absence of oxygen*, Sn is *anodic* to steel (plating of steel cans for food ('tin' cans)) and thus tin will corrode first. The Sn-salts that form are non-toxic. However, in the *presence of oxygen* Sn acts *cathodic* (Figure 8.9).

- **Intergranular corrosion**
  One example is shown in Figure 8.10, where precipitation of a β-phase in an α-matrix occurred. Such a two-phase microstructure can provide micron-scale galvanic cells at phase boundaries (note the correlation to phase diagrams), due to local microstructural inhomogeneities.

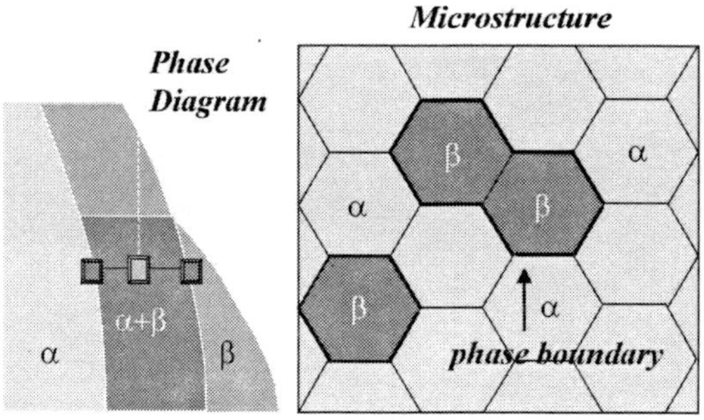

Fig. 8.10 Schematic showing the correlation between intergranular corrosion and the two-phase field in a binary phase diagram.

- **Pitting corrosion**
  Pitting corrosion (Figure 8.11) is caused by the formation of local sites of anodes and cathodes due to microstructural variations such as grain boundaries or dislocations on the metal surface. Electrochemical reactions usually occur at such surface sites, since atoms at grain boundaries or dislocations are more disordered (dangling bonds) than in the interior of the single crystal. Therefore, they are chemically more active (Figure 8.11).

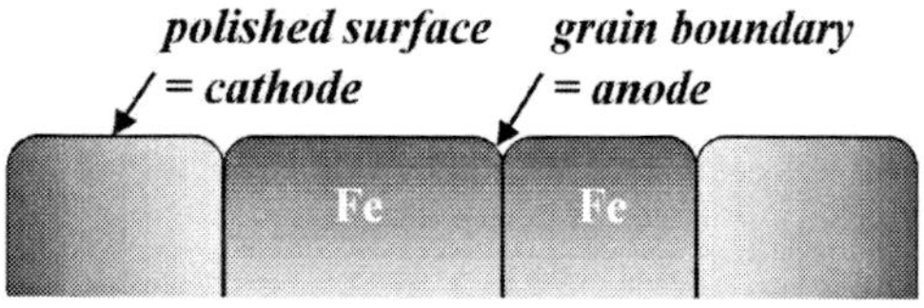

Fig. 8.11 Schematic illustrating the process of pitting corrosion at microstructure variations such as grain boundaries or dislocations.

- **Crevice corrosion**
  When for example water is trapped in a crevice of the surface and the aggressive reaction products (HCl, H_2SO_4) that form during corrosion stay also trapped in the crevice region.

- **Stress corrosion**
  Stress corrosion is a combined effect of stress (residual or applied) and specific corrosion environment, causing **slow crack growth**. This is very dangerous, if it occurs in large structures since the stress-intensity factor, K_I, is proportional to the square root of crack length. When the crack length increases (Figure 8.12), K_I also increases and catastrophic failure can occur.

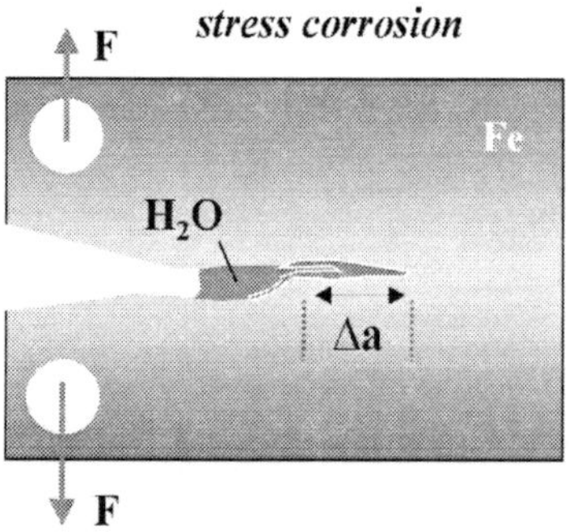

Fig. 8.12 Schematic illustrating the correlation between stress corrosion and change in crack length, Δa.

- **Erosion corrosion**
  Mechanical wear aids in the formation of local damaged areas on the surface. If the surface layer, which may be a protective oxide such as Al_2O_3 on Al, is damaged, corrosion will proceed upon erosion (Figure 8.13).

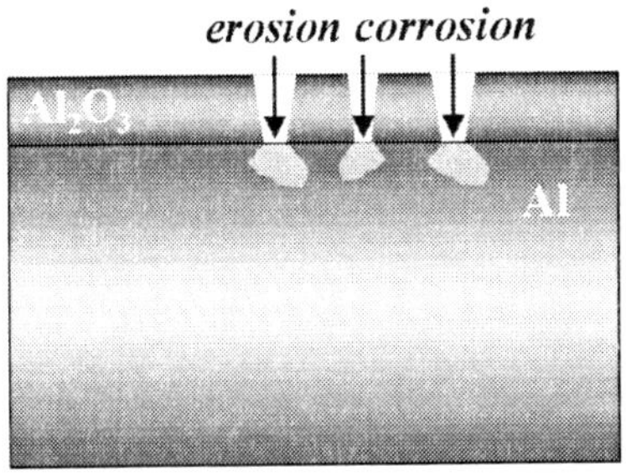

Fig. 8.13 Schematic showing effect of surface wear on erosion corrosion.

There are three types of **protective coatings** that can be applied:

- **metallic** coatings as for example Zn on steel or Cr-Ni on steel (Cr-plated bumpers). ***Note*** that in the latter case, the Ni is the metal that protects steel from corrosion while Cr is mainly for appearance (Ni is less shiny).
- **inorganic** coatings typically of ceramics or glasses (porcelain enamels). The main problem here is the large difference in thermal expansion coefficient, which leads to cracking and spalling of the protective surface coat.
- **organic** coatings such as paints and varnishes.

8.2.2 Corrosion Protection

There are a number of ways, to effectively **protect** a metal surface from corrosion, as shown in Figure 8.14. Apart from **materials selection**, where a more corrosion resistant material replaces a less noble metal (which may be too costly), **design** and **environmental control** lower the probability of aggressive corrosion media to be in contact with the metallic component.

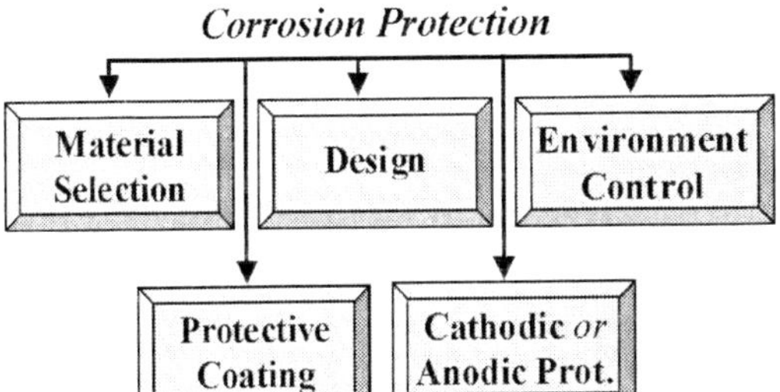

Fig. 8.14 Schematic showing the different approaches of corrosion protection.

Another very effective way to prevent metal corrosion is the **cathodic** and **anodic protection**. If we consider the corrosion of large iron components (ship hull), the basic corrosion reaction can be written as:

$$Fe^0 \Longrightarrow Fe^{2+} + 2e^-, \text{ anodic}$$

$$2H^+ + 2e^- \Longrightarrow H_{2(g)}, \text{ cathodic.}$$

However, when sufficient electrons are externally supplied, the anodic reaction will be suppressed and, in turn, corrosion will not occur. The addition of an **artificial anode** prevents the iron to corrode (Figure 8.15).

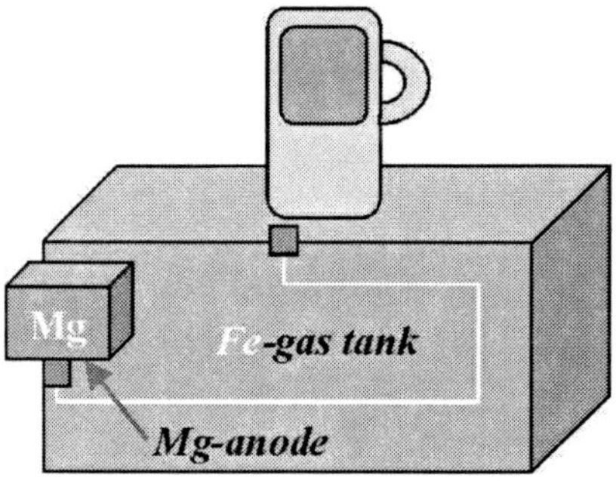

Fig. 8.15 Schematic showing the use of an artificial Mg-anode to prevent corrosion of an Fe-gas tank.

The utilization of a **cathodic protection** system was developed in the early 1950s to also prevent underwater corrosion of ship hulls. An **external voltage** is used to *oppose* the one due to the electrochemical reaction. The impressed voltage stops the flow of electrons required for the corrosion reaction to proceed.

Note that the corrosion rate, R_{corr}, can be calculated from the corresponding current density, I, using:

$$R_{corr} = \frac{I}{n \cdot F} \qquad (8.4)$$

where n is the number of electrons transferred during the corrosion reaction and F is the Faraday constant (96,000 C/mol).

Questions

What does the Pilling-Bedworth ratio, *R*, describe?

What is the oxidation kinetics of a spalling surface coating versus an adhering oxide coat?

Where does oxidation (galvanic cell) take place; at the anode or cathode?

How can you protect a large plain steel container from severe corrosion attack?

9 Ceramics and Glasses

Ceramics and glasses represent one of the major classes of engineering materials. The word *ceramics* has its origin in the Greek work *keramos*, which means "burnt material". Early discoveries of ceramics and glasses date back to 3500 B.C. and there are still a number of historical sites preserved (pyramid of Giza, Parthenon temple) which illustrate the durability of ceramics.

Crystalline ceramics include traditional silicates (refractories, porcelain, cement) and are widely used. Since oxygen and silicon together account for about 75% of the elements of the earth crust (Figure 9.1), natural silicates are abundant and commonly used as resources for ceramic processing.

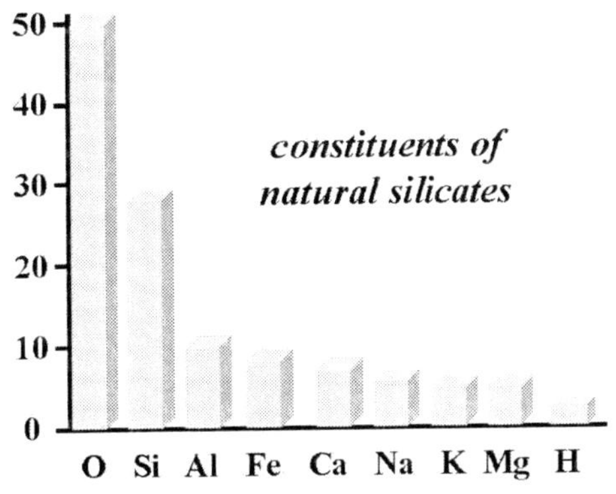

Fig. 9.1 Relative abundance of elements in the earth crust that are constituents of various silicates.

Clay for example can easily be shaped by adding water. Upon drying and subsequent firing, the mechanical properties of the silicate compound improves drastically.

The **definition** of a ceramic material is a compound that is composed of *metallic* and *non-metallic* elements and is characterized by *ionic* and/or *covalent* bonding.

In particular due to their specific bonding character, ceramic materials reveal an interesting combination of various favorable properties such as:

- **corrosion resistance**
- **high-temperature resistance**
- **wear resistance**
- **high strength**
- **high creep resistance**
- **lightweight**.

In contrast to crystalline ceramics, **glasses** only show short-range order on the atomic level and are therefore amorphous solids. The absence of crystallinity, which is a result of their chemical composition and specific processing techniques (fast cooling rate) gives glasses a unique set of mechanical and optical properties.

Glass ceramic, on the other hand, is another type of predominantly crystalline material, which is initially formed as a glass and then carefully crystallized in order to gain the wanted crystalline phases and corresponding properties. One example is the crystallization of Li-Al-silicates, which commonly show a low thermal expansion coefficient leading to thermal shock resistant glass ceramics.

9.1 Crystalline Ceramics

The crystal structures of ceramics are more complex than those of metals. *Strong atomic bonding* (ionic and/or covalent) leads to atoms on specific sites in the unit cell and results in the high melting temperatures, high stiffness, and high strength of ceramics. There are no random solid solutions rather than the possibility of ion mixing on the cation or anion sublattice. Since ceramics are generally defined as the combination of a metal-cation with a non-metallic anion, a wide variety of different ceramics are known. Commercial products such as **whitewares** (pottery, translucent porcelain) are heat-treated ceramics of white color and have a rather fine-grained microstructure. Their basic components are:

- Clay ($2Al_2O_3{\cdot}4SiO_2{\cdot}4H_2O$) ("kaolinite") $Al_4(OH)_8Si_4O_{10}$
- Silica (SiO_2) ("flint")
- Feldspar ($K_2O{\cdot}Al_2O_3{\cdot}6SiO_2$).

Clay, a hydrated alumino-silicate, is the basis for many ceramic components such as building bricks, sewer pipes, drain tiles, roofing tiles, floor tiles, dinnerware, and dental porcelain.

Refractories, also based on fired clay, are high-temperature stable structural ceramics that play a crucial role in industrial applications (steel making).

There are also a number of **non-silicate** based ceramics, which are very important as either structural or electronic and magnetic materials. A few examples include:

- **oxides**: Al_2O_3, MgO, SiO_2, $Al_6Si_2O_{13}$ (mullite), $MgAl_2O_4$ (spinel), Al_2TiO_5, ZrO_2 (commonly stabilized)
- **nitrides**: Si_3N_4, BN, TiN, AlN (high thermal conductivity, used in electronic packaging)
- **carbides**: SiC, WC, TaC, B_4C (ceramic armor), TiC (wear resistant coatings)
- **electronic ceramics**: $BaTiO_3$, $SrTiO_3$ (based on the perovskite structure)
- **magnetic ceramics**: $NiFe_2O_4$, $NiAl_2O_4$ (based on the spinel structure).

Some of these ceramics have been known for several decades such as **SiC**, which is widely used as a heating element (gas igniter) or abrasives (grinding, polishing).

Silicon nitride, $\mathbf{Si_3N_4}$, and the related material group of SiAlONs (a solid solution between Si_3N_4 and Al_2O_3 with mixing on *both sublattices*) have been the major components in worldwide gas-turbine research (automotive application) over the last three decades (see also below).

Alumina, $\mathbf{Al_2O_3}$, is often used for refractory tubing, high purity crucibles, spark plug insulators, and electronics substrates.. Al_2O_3 is also utilized in high quality electrical applications, where low dielectric loss and high electrical resistivity is needed.

Densification of alumina components typically involves **sintering aids** such as MgO (1-2 wt%) or SiO_2 (2-5 wt%), which promote sintering via grain-boundary diffusion and also control Al_2O_3 grain size (Figure 9.2).

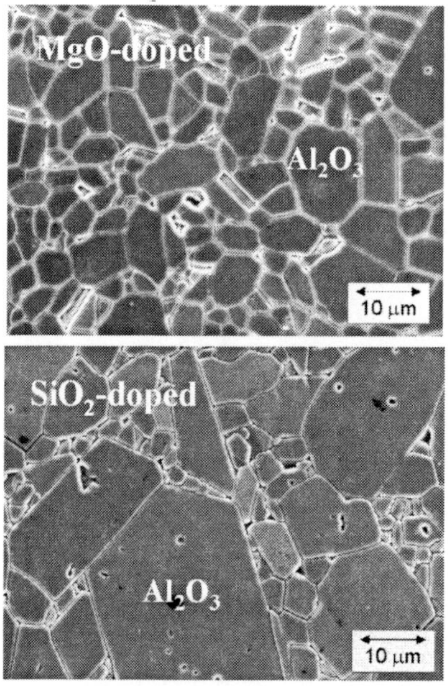

Fig. 9.2 SEM images of Al_2O_3 microstructures densified with the addition of MgO (top) and SiO_2 (bottom).

Silicon carbide, **SiC**, is employed in wear resistant parts such as ball bearings, extrusion dies, rocket nozzles, heat exchanger tubes, or mixers for ceramic powders. SiC is also used as fibers in composite materials due to its high-temperature oxidation resistance, however, note that SiC fibers are inherently brittle.

SiC ceramics are commonly processed by reacting graphite with molten Si. As a result of this process, free Si (8-20 wt%) is present in this **SiSiC** composite, which is shown in Figure 9.3.

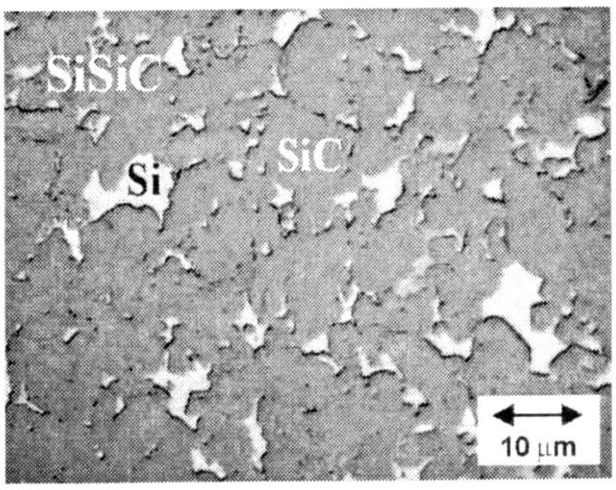

Fig. 9.3 SEM image of a SiSiC composite; the bright areas are residual Si-metal while the gray regions are SiC.

When starting from SiC powders, typically about 1 wt% of B and/or Al is added to the SiC powder to enhance densification. *Note* that, although this process has been used for more than 40 years, the details of the densification mechanism are still unknown.

Silicon nitride, $\mathbf{Si_3N_4}$, is used as engine components, particularly, for diesel engines, or as turbochargers for spark-ignited engines and ball bearings for aerospace applications (low temperatures). Si_3N_4 has high strength and relatively high fracture toughness (in-situ toughening); which is why it was also referred to as "*steel of ceramics*".

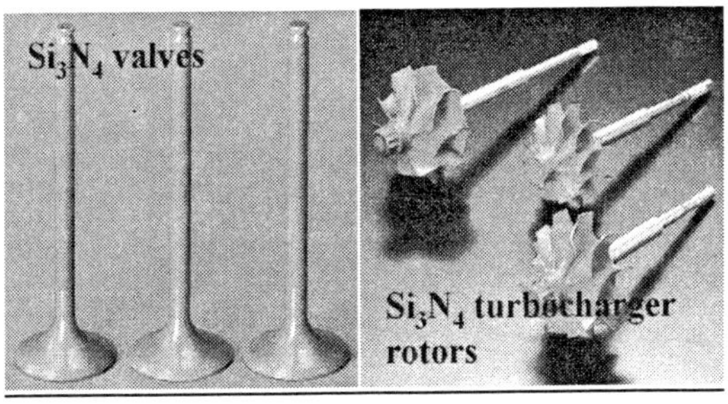

Fig. 9.4 Si_3N_4 parts (valves and turbocharger rotors) for potential use in automotive (diesel) engines.

Si_3N_4 has, in general, a very favorable combination of properties such as light-weight, high strength, reasonable toughness, and high refractoriness. However, since Si_3N_4 *dissociates incongruently* above 1800 °C (decomposition into liquid Si and gaseous N_2), processing is more of a challenge. Typically, 1-5 wt% of MgO, Al_2O_3, and/or Y_2O_3 is added as a sintering aid to promote densification at lower temperatures via the "classical" liquid-phase sintering process.

9.1.1 Mechanical Properties

Apart from the advantageous properties mentioned earlier, ceramics are **inherently brittle**. This means that ceramics, revealing high strength and stiffness, show little or no plastic deformation when external stress is applied. As a consequence, a small crack can easily propagate within a ceramic component. Hence, small processing flaws such as residual pores can act as crack initiation sites and cause catastrophic failure. This brittleness is the major limitation when ceramic components are being used.

One approach to overcome this obstacle is careful processing to **reduce** the maximum **flaw size** (correlation between K_I and a).

Another way to introduce ceramic parts in industrial applications is to properly **design** the component so that the ceramic part is loaded under **compression**.

Ceramics are much stronger in *compression* than in tension (~8 times), because cracks close under compression. For metals, failure occurs by *yielding*, which happens at nearly equal loads in tension and compression, as shown in Figure 9.5.

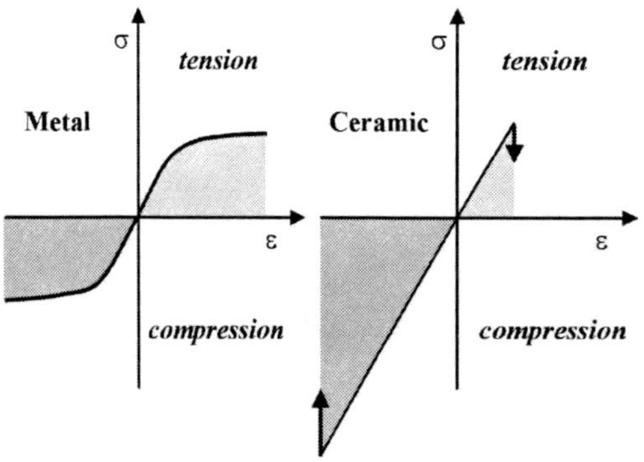

Fig. 9.5 Stress-strain curves (tension vs. compression for metals and ceramics). *Note* that ceramics are much stronger under compressive loading.

The stress-strain curve of ceramics in Figure 9.5 reveals that crack propagation occurs before yielding becomes effective. For ceramics no yield stress, σ_Y, can be defined; instead, they show a linear stress-strain behavior up to the stress level where catastrophic failure occurs. As a result, ceramics have **low fracture toughness** values:

- *most* **ceramics** **K_{IC} < 4 MPa(m)$^{1/2}$**
- high tech ceramics 5-10 MPa(m)$^{1/2}$
- ceramic composites 8-20 MPa(m)$^{1/2}$

- *most* **metals** **K_{IC} >100 MPa(m)$^{1/2}$**
- aluminum 26 MPa(m)$^{1/2}$
- beryllium 9-13 MPa(m)$^{1/2}$

The strength of ceramic materials can easily be determined by the **modulus of rupture** (MOR) test (no need for the complex tensile test). The strength value is calculated from a relatively simple bending test and is independent of the geometry of the given sample. Figure 9.6 shows the geometry of the MOR test (for 3-point bending).

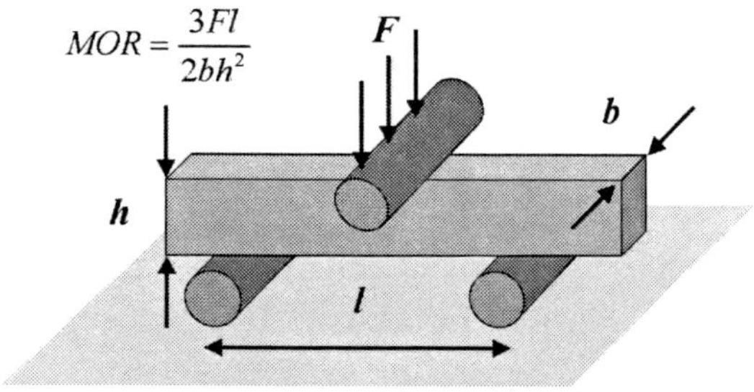

Fig. 9.6 Schematic showing the setup for a modulus of rupture (MOR) test in 3-point bending.

The corresponding strength value of the ceramic can easily be determined from equation 9.1:

$$MOR = \frac{3Fl}{2bh^2} \qquad (9.1)$$

where F is the load at failure, b and h are related to the dimensions of the test sample, and l is the distance between the two lower fixtures, as shown in Figure 9.6.

The MOR value is very similar in magnitude to the tensile strength of the ceramic, since the failure mode in the bending test is also **tensile** (at least at the outermost edge of the sample). This bending test is typically easier to perform for brittle materials as compared to a standard tensile test used for metals.

Note that the Poisson's ratio of ceramics is commonly at the lower end of the narrow range (0.24-0.35), ~0.25, while metals and alloys are close to the upper level, ~0.33.

9.1.2 Toughening of Ceramics

Griffith theory of fracture works reasonably well for brittle materials. When comparing the stress-strain curves in Figure 9.5 (metal vs. ceramic), it becomes obvious that for a ceramic material, the energy required to fracture the sample and to create two new surfaces (surface energy) is lower than the energy required to gain further elastic deformation.

This brittle behavior of ceramics has triggered extensive research into possible mechanisms to improve their **resistance against crack propagation** (increase their fracture toughness). In this section we will briefly discuss two toughening mechanisms in **monolithic ceramics**: (i) in-situ toughening of Si_3N_4, and (ii) transformation toughening of partially stabilized ZrO_2.

In-situ toughening of $\mathbf{Si_3N_4}$ ceramics is obtained by growing large elongated Si_3N_4 grains in a fine-grained Si_3N_4 matrix, as shown in Figure 9.7.

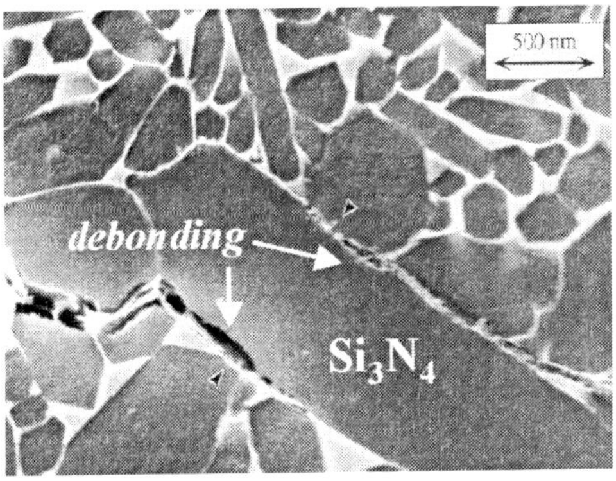

Fig. 9.7 SEM image of in-situ toughened Si_3N_4. Note that debonding occurs at the interface between the large grain (elastic bridge) and the matrix grains (weak interface).

The large grains embedded in the matrix act as elastic bridges in the crack wake upon crack propagation, which makes it more difficult for the crack to extend. It is important to note that a prerequisite for the mechanism of in-situ toughening to be operative is that **debonding** along interfaces occurs (we will encounter the aspect of interface debonding again when talking about composites). If strong interface bonding is present, the crack would fracture the elongated grains transgranularly and no effect on toughness would be noticed. Apart from elastic bridges in the wake of the crack, the crack path also becomes longer and more energy is dissipated. Both mechanisms contribute to the increase in fracture toughness. ***Note*** that Si_3N_4 with strong interface bonding reveals a fracture toughness of ~2.5 $MPa(m)^{1/2}$ while a K_{IC} of ~7-8 $MPa(m)^{1/2}$ was reported for in-situ toughened Si_3N_4.

Transformation toughening of $\mathbf{ZrO_2}$ is a classic case where ceramic research has focused over many years to design special microstructures that can effectively resist crack propagation.

There are **three polymorphs** of ZrO_2 (identical composition but different crystal structure). The exact temperature where the respective polymorphs are stable depends on the purity (dopant level) of the polycrystal; for undoped ZrO_2 the temperatures are as follows:

- **monoclinic** **RT – 1170 °C**
- **tetragonal** **1170 – 2370 °C**
- **cubic** **>2370 – 2600 °C**.

The room temperature stable phase is monoclinic ZrO_2; however, tetragonal and cubic ZrO_2 can be stabilized at lower temperatures by adding a stabilizer (MgO, CaO, CeO_2, or Y_2O_3). The high-temperature stable cubic phase can then exist even at room temperature. The phase diagram given in Figure 9.8 shows the stabilization of cubic ZrO_2 down to low temperatures. ***Note*** that synthetic gemstones similar to diamond are often stabilized cubic-ZrO_2, which is called **fully stabilized zirconia** (FSZ).

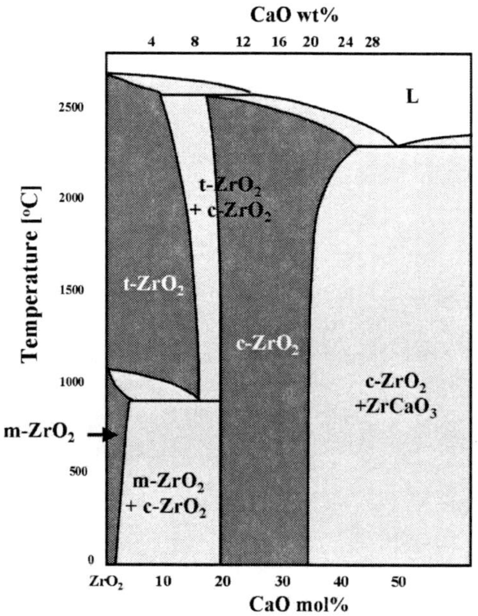

Fig. 9.8 Phase diagram of CaO-doped ZrO_2. *Note* that c-ZrO_2 can be stabilized down to RT.

In order to activate the mechanism of transformation toughening, one has to design a microstructure where the tetragonal to monoclinic **phase transformation** locally takes place. The recipe to yield such a microstructure is given below (see also the corresponding phase diagram in Figure 9.8).

1. dope ZrO_2 with about 18 mol% CaO and heat up to 1800 °C where the cubic solid solution forms
2. cool quickly to room temperature before any transformation takes place
3. heat to 1400 °C to let tetragonal ZrO_2 precipitates form
4. fast cooling to retain the microstructure of tetragonal ZrO_2 precipitates finely dispersed in a cubic ZrO_2.

Note that if the t-ZrO_2 precipitates get too big, they would spontaneously transform to the monoclinic phase upon cooling. Therefore, the heat treatment at 1400 °C must be well controlled. The resulting material is termed **partially stabilized zirconia** (PSZ). A schematic of the PSZ microstructure is given in Figure 9.9.

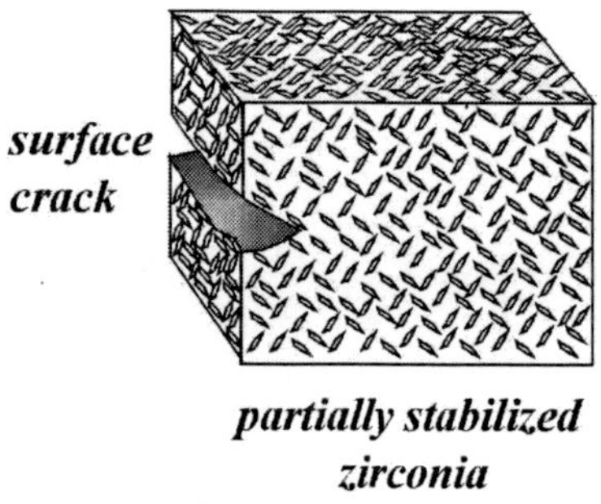

Fig. 9.9 Schematic of the microstructure of PSZ (cubic matrix is white, tetragonal precipitates are gray); compare also to Figure 9.10.

The tetragonal phase is metastable at room temperature. When a small crack propagates, the stress field at the crack tip causes a **martensitic phase transformation** to monoclinic ZrO_2. This transformation is associated with a **volume expansion** of approximately 3 vol%. The excess volume at the crack tip and at the crack wake opposes crack propagation and closes the crack (Figure 9.10). Consequently, it becomes increasingly difficult to propagate the crack resulting in a marked improvement in fracture toughness with K_{IC} values for PSZ of >10 MPa(m)$^{1/2}$.

Note that the martensitic phase transformation is a diffusion-less (displacive) process and, therefore, will always occur.

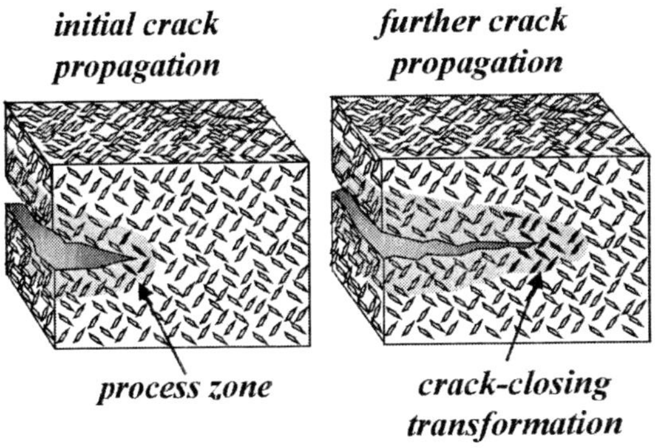

Fig. 9.10 Schematic of the mechanism of transformation toughening operative in PSZ. Martensitic transformation of t-ZrO_2 to m-ZrO_2 (volume increase) causes crack closure.

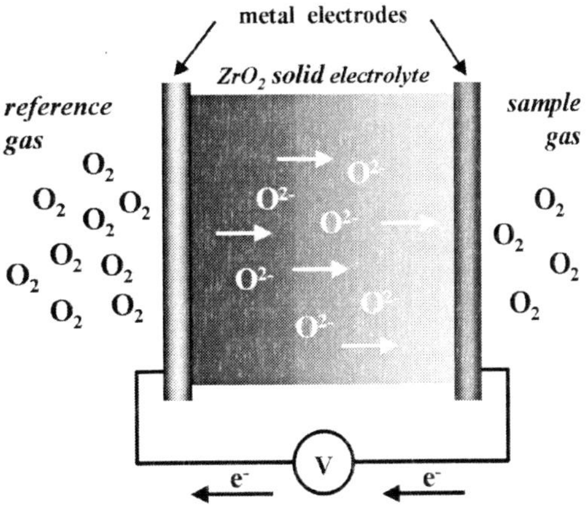

Fig. 9.11 Schematic of an oxygen sensor (CaO-doped ZrO_2).

Due to its high fracture toughness and very high strength (maximum values reported are 1800 MPa) ZrO_2 is often used as hammers, putters, or **knives** and **scissors**, since ZrO_2 ceramics retain their sharp edge for a long time; however, they are still brittle.

ZrO_2 ceramics are also used as **oxygen sensors** and play an important role in fuel cell applications. Here, the zirconia acts as a solid electrolyte with high ion conductivity. In the example of an oxygen sensor, CaO-doped ZrO_2 is used. CaO doping leads to the formation of oxygen vacancies, as indicated by the **Kröger-Vink notation** of the doping process:

$$CaO \xrightarrow{ZrO_2} Ca''_{Zr} + O_O^x + V_O^{\bullet\bullet} \quad (9.2)$$

Ca^{2+}-ions replace Zr^{4+}-ions on their lattice sites. The oxygen sublattice is not affected (no exchange) with the exception of the formation of **oxygen vacancies** (charge compensation for Ca^{2+} incorporation). These vacancies enable the oxygen-ion conductivity needed in the oxygen sensor (Figure 9.11).

The electrons flow from the electrode with a low oxygen partial pressure, p_{O_2}, to the one with a high p_{O_2}. The oxygen of the reference gas is converted to O^{2-}-ions (reduction), which will pass through the solid electrolyte.

$$\tfrac{1}{4}\,O_{2(g)} + e- \Rightarrow \tfrac{1}{2}\,O^{2-}_{(g)} \quad (9.3)$$

The voltage measured between the two metal electrodes is proportional to the oxygen partial pressure of the sample gas.

9.1.3 Processing of Ceramics

Processing of ceramics (how to make ceramic components) can basically be divided into three major processing steps:

- **powder preparation**
- **forming** (shaping)
- **densification** (thermal treatment).

Ceramic components made of Al_2O_3, ZrO_2, or SiC are, in general, rather **expensive**, because processing requires high purity oxide or non-oxide starting powders, which are costly to synthesize. Moreover, the heat treatment (sintering) and final surface finish (grinding and polishing) add to the overall costs of ceramic components.

One important initial step during ceramic processing is the homogeneous **mixing** of the starting powders, which may differ in size, shape, and composition. The starting powders can be blended in the wet or dry state. **Lubricants** or **binders** are frequently added, which helps to yield a dense powder compact (green body) that can be shaped and which holds together. ***Note*** that the organic binders have to be carefully burned out before final densification (typically at 400 °C). For many traditional ceramics (refractories), water is often used to blend the respective ingredients.

Forming of ceramic powders commonly employs cold shaping rather than hot forming. The forming process can be performed under dry, plastic, or liquid conditions:

- **extrusion**
- **slip casting**
- **injection molding**
- **dry pressing.**

The shaping technique of **slip casting** can be described by the following steps, where steps 1-4 correspond to the actual forming process:

1. preparation of a **liquid/powder mixture** called a slip, which is a stable suspension
2. the slip is poured into a **mold** (plaster of paris) where some liquid is absorbed by the mold so that a semi-hard layer is formed against the mold wall
3. when the semi-hard layer is thick enough, the **excess slip** is **poured out** (drain casting)
4. the **cast** is allowed to **dry** in the mold and then removed having its final shape
5. **heat treatment** of the cast part at high temperature; *note* that during heat treatment shrinkage will occur.

During **extrusion**, the ceramic powder-binder mixture is forced through a die, which can produce only simple shapes such as single cross sections or tubes. This forming process is used for refractory bricks, sewer pipes, hollow tiles, or electrical insulators.

There are varies types of **pressing** ceramic powders:

- ***dry pressing*** is used extensively and is a *uniaxial* pressing technique with small amounts of **water** and/or **organic binder** in the die, which then requires an additional firing step
- ***isostatic pressing*** places ceramic powders in *airtight* flexible containers (usually rubber) which are placed inside a chamber of *hydraulic fluid* to which pressure is applied; since binder is commonly used, the parts also have to be fired
- ***hot pressing*** is a combined pressing and firing processing technique where both *uniaxial* and *isostatic* pressing at high temperature is possible; the latter is then called *hot-isostatic pressing* (HIPing).

Sintering of a pre-shaped ceramic green body, with the organic binder already removed, refers to the process of **firing** and **consolidation** to form a strong, dense component. During sintering, pores are removed, which is accompanied by shrinkage of the ceramic part and the formation of strong bonds between adjacent particles.

Note that the overall driving force of the sintering process is the reduction of surface area (lowering surface energy). The various paths of matter transport during the initial stage of sintering are shown in Figure 9.12.

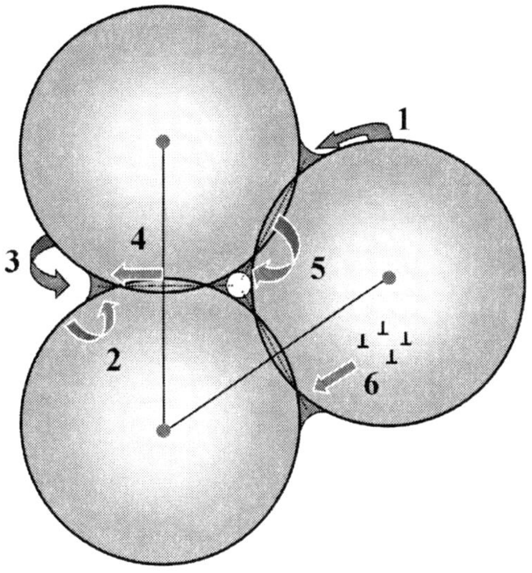

Fig. 9.12 Different paths of matter transport during the initial stage of sintering.

The **initial stage** of sintering involves **rearrangement** of powder particles and the formation of a **neck** between them. The various transport mechanisms, as shown in Figure 9.12, only partly contribute to the densification of the powder compact, depending on if the transported mass originates at the surface or from the volume of the particles.

Surface diffusion (#1) promotes the growth of the neck region, but no densification occurs (no change in particle spacing).

Volume diffusion (bulk mass transport; #2), that also originates at the particle surface and contributes to neck growth, does not densify the powder compact.

Vapor transport (evaporation and condensation; #3) has a similar effect as surface diffusion. No densification occurs, since the actual mass flow originates and terminates on the particle surface.

Grain-boundary diffusion (#4) is the most important transport mechanism with respect to densification of ceramic components.

Volume diffusion (#5), originating at grain boundaries, is commonly more sluggish and depends on the defect concentration of the particles.

Similarly, **bulk diffusion** (#6), originating at dislocations, is less favorable as compared to grain-boundary diffusion.

Only bulk diffusion processes such as #4, #5 and #6 actually contribute to shrinkage (densification), *since* the mass transport originates in the interior of the particles and is deposited at the neck region.

During the **intermediate stage** of sintering, where the neck regions between particles grow, residual porosity is substantially decreased, leading to **shrinkage** of the component.

In the **final stage** of sintering, the initially open pores become closed and are slowly eliminated, generally by **vacancy diffusion** from pores along grain boundaries. In addition, grain size increases during the final sintering stage (depending on temperature and sintering time).

Note that densification of ceramic powders with high melting or decomposition temperatures (MgO, Al_2O_3, Si_3N_4, SiC) often requires the addition of small volume fractions of oxides in order to lower the sintering temperature. This process is called **liquid-phase sintering**.

9.2 Amorphous Ceramics

Non-crystalline ceramics, i.e., glasses, are commonly based on the SiO_2 network. The basic building unit for glasses is again the SiO_4^{4-}-tetrahedron. In pure silica glass (fused silica) all silica tetrahedra (Figure 9.13) share corners with each other forming a 3-dimensional complete network (Figure 9.14).

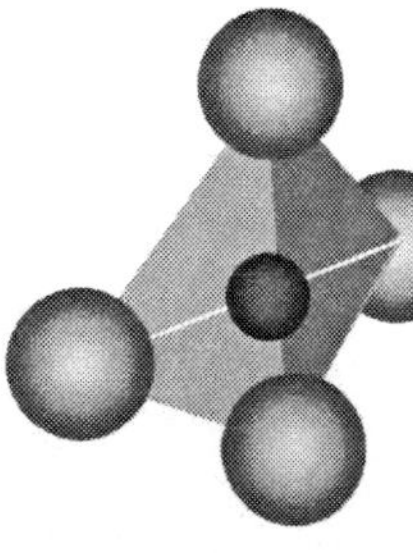

Fig. 9.13 The basic building unit of most glasses is the SiO_4^{4-}-tetrahedron.

Glasses are processed from a melt, which is cooled rapidly and transforms into a supercooled solid with an amorphous network structure (Figure 9.14). This model of a random glass network structure was first introduced by W.H. Zachariasen in 1932.

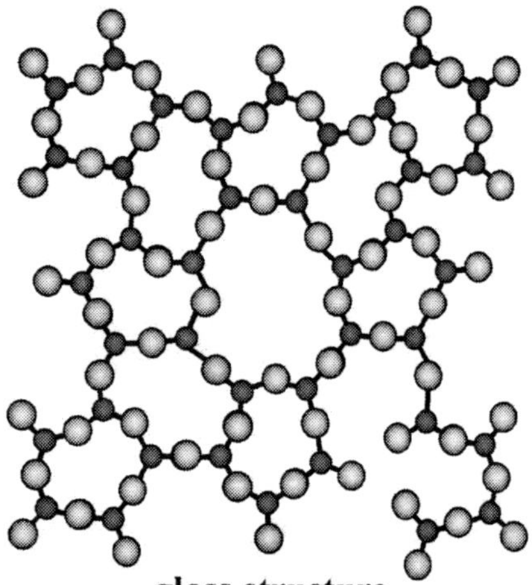

Fig. 9.14 3-dimensional glass structure of fused silica (Zachariasen model).

The addition of other oxides such as Na_2O, K_2O, or Al_2O_3 changes the structure of the glass network. Incorporating, for example, Na_2O into pure silica glass increases the O/Si-ratio and, as a consequence, breaks up a fraction of the bridging oxygen, as shown in Figure 9.15.

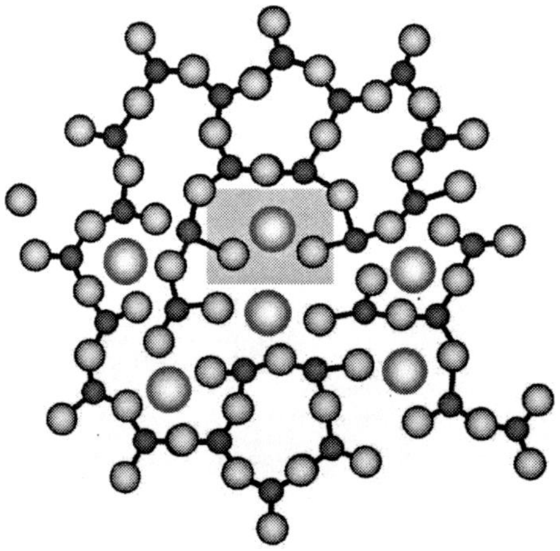

Fig. 9.15 Random Na_2O-SiO_2 glass network. Note that some of the bridging oxygens are disrupted; the addition of one Na_2O molecule is indicated by the boxed area.

One distinguishes between different types of additives. Similar to silica (SiO_2), B_2O_3, GeO_2, and P_2O_5 are considered **network formers**, since they can build a glass network on their own. Additions such as Na_2O, K_2O, CaO, MgO, BaO, or PbO are termed **network modifiers**, because they disrupt the glass network, as shown in Figure 9.15.

There is a third group of additives that are termed **intermediates**, for example Al_2O_3, TiO_2, or ZrO_2 that do not promote or disrupt the glass network, but substitute in the structure. Intermediate oxides cannot form a glass network by themselves. Al_2O_3 is the classic example of an intermediate addition since it can replace Si^{4+} in the network.

Note that the addition of TiO_2 tends to promote local crystallization within the glass (used in processing of glass ceramics).

Tab. 9.1 Composition of some industrially important glasses.

	Composition [wt%]					
	SiO_2	B_2O_3	Al_2O_3	Na_2O	CaO	MgO
Fused silica	100	-	-	-	-	-
Boro-silicate	76	13	4	5	1	-
Window glass	72	-	1	13	11	3
Container glass	73	-	2	14	10	-
Fiber (E) glass	54	8	15	-	22	-

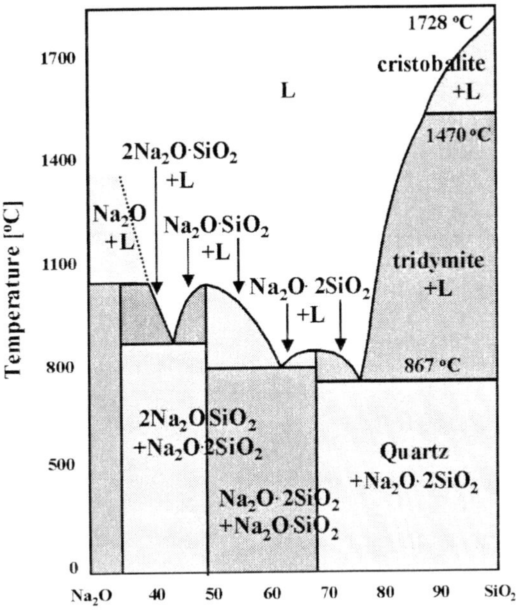

Fig. 9.16 Na_2O-SiO_2 phase diagram. *Note* the pronounced drop of the SiO_2 melting temperature even with small additions of Na_2O (network modifier).

When various network formers are added to the glass melt, the network structure changes and, simultaneously, **glass properties** dramatically change. For example, the glass **melting temperature**, T_m, is substantially reduced and glass **viscosity**, η, is markedly lowered. Lowering of T_m with the addition of glass modifiers (Na_2O) can also be seen in the Na_2O-SiO_2 phase diagram (low eutectic temperature), given in Figure 9.16. This change in properties is a consequence of the disrupted glass network.

The main advantage of lowering the glass melting temperature and viscosity is the ease of **glass forming**. However, depending on the glass composition, **chemical durability** may be lost. For example, sodium silicate glasses are water soluble ("waterglass"), while the use of additional CaO results in **soda-lime glasses** that are insoluble in water and show overall an improved chemical durability.

When plotting **specific volume**, V_s (volume per unit mass), versus temperature for a **single crystal** (Figure 9.17), a linear increase in V_s with raising temperature up to the melting temperature, T_m, is found. At T_m, a discontinuity in V_s increase occurs. Above T_m, the increase in specific volume with increasing temperature again follows a linear relationship.

A **glass**, however, behaves differently. When cooling down from a temperature above T_m, a linear decrease in V_s is observed even below the T_m of the crystalline phase. At one temperature, defined as the **glass transition temperature**, T_g, there is a more gradual deviation from linearity. Below T_g, the supercooled liquid becomes a solid where again specific volume decreases linearly with decreasing temperature. The temperature region where it is best to form a given glass is between T_g and T_m and is termed the **glass working region**. In this region, a glass behaves like a rather "thick" liquid (honey) and can be formed for example by glass blowing.

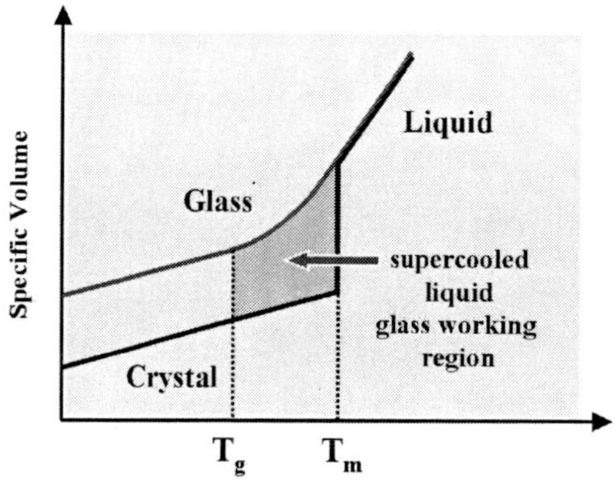

Fig. 9.17 Specific volume vs. temperature for a glass and a single crystal.

The viscous behavior of a supercooled liquid can be quantitatively described by the glass viscosity, η, which is the proportionality constant between a shear force, F/A, and the velocity gradient, dv/dx:

$$\tau = \frac{F}{A} = \eta \cdot \frac{dv}{dx} \tag{9.4}$$

Note that the glass molecules move at any temperature, which allows even at room temperature glass to slowly flow, though on a more "geological" time scale.

The main reason to **strengthen** glass is its inherently low fracture resistance and its intrinsic brittleness. The basic strengthening concept is to induce **residual compressive stresses** at the glass **surface** (Figure 9.18). There are two methods used to introduce such a surface stress state:

- **Cooling/tempering**

  The glass surface is strengthened via **rapid cooling** (air spray). The surface cools first (low thermal conductivity of glass) and it contracts while the interior is still warm and can accommodate shape change via viscous flow. When the outside is cooled below T_g, it will act like a rigid solid, while the interior further contracts. Hence, the center of the glass component tries to force the surface to contract, while the surface tries to force the center to expand. This results in internal tensile stresses and compressive stresses at the surface, leading to a higher bending strength and an improved resistance against surface crack propagation in the glass.

- **Chemical strengthening**

  Large cations like K^+ are incorporated into the outer surface area, for example by liquid ion exchange (immersing the glass into a glass melt of different composition). Replacing Na^+ by the larger K^+-ion also causes compressive stresses at the glass surface.

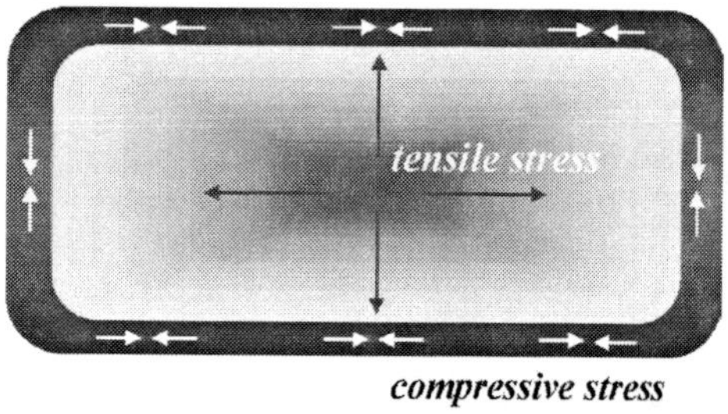

Fig. 9.18 Compressive stress at the glass surface strengthens the glass and increases its resistance against surface crack propagation.

Note that glass also undergoes **stress corrosion.** When for example water is present at the crack tip, following reaction occurs: $SiO_2 + 2H_2O \Rightarrow Si(OH)_4$, breaking up the SiO_2 network and simultaneously increasing the initial crack length.

9.3 Glass Ceramics

In contrast to what the term "glass ceramic" suggests, glass ceramic products are highly **crystalline**; up to 90-95 vol%. However, processing of glass ceramics starts with a melt, which is then slowly cooled in a controlled way, in order to crystallize the wanted phases and to tailor the properties of the glass ceramic component.

Glass ceramics typically reveal low porosity, rather small crystallites (surrounded by residual glass), and reveal good mechanical behavior. Many glass ceramics are based on the **Li_2O-Al_2O_3-SiO_2** system, which allows the crystallization of Li-Al-silicates with *low thermal expansion* coefficients. ***Note*** that **β-eucryptite**, $LiAlSiO_4$, has a *negative* thermal expansion and, therefore, allows the production of glasses that are thermal-shock resistant (see also Chapter 12 on Thermal Properties).

Comparing glasses with glass ceramics, a common viewpoint can be seen: **nucleation** of crystalline phases. When processing glasses via rapid cooling, one tries to *prevent* crystallization, while the production of glass ceramics involves a careful heat treatment to *initiate* crystallization.

Crystallization can be conveniently separated into two physical processes: (i) nucleation of crystallites and (ii) crystallite growth. Formation of a *stable nucleus* is favorable, when the decrease in free volume energy (crystals commonly show higher density than the corresponding liquid) overcomes the increase in surface energy (new surface is formed during nucleation). The change in Gibbs free energy for a spherical nucleus of radius r can be expressed as:

$$\Delta G(r) = 4\pi r^2 \gamma_{sl} + \frac{4}{3}\pi r^3 \Delta G_V \quad (9.5)$$

where γ_{SL} is the solid-liquid surface energy and ΔG_V is the volume free energy change of crystallization. The two terms, **surface energy change** and **volume free energy change** during nucleation, result in a net change in Gibbs free energy depending on the radius of the nucleus (Figure 9.19)

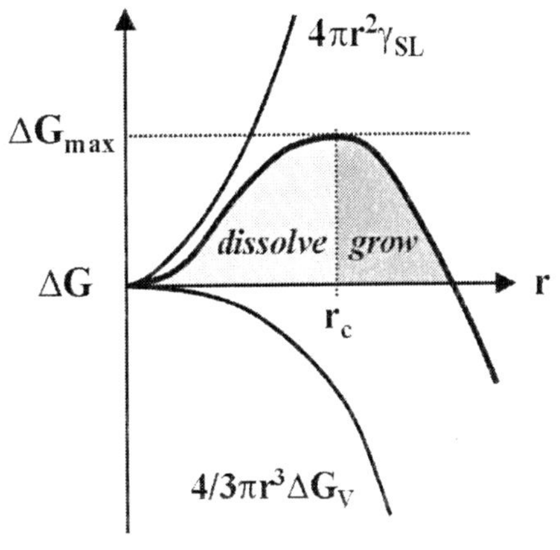

Fig. 9.19 Gibbs free energy vs. nucleus radius. *Note* that below the critical radius, r_c, the nucleus will dissolve while above r_c it will grow.

Figure 9.19 reveals that crystallization (and growth) only occurs when the nucleus radius exceeds r_c (glass ceramics). Below that critical value, the initially formed nuclei will dissolve, resulting in an amorphous glass structure.

Here, we will briefly introduce the **TTT-diagrams**, which can be used to describe the time dependent reactions that occur during heat treatment. The three T's stand for temperature, time and transformation. Such diagrams reveal the **kinetics** of reactions, which are referred to as time-dependent phase transformations.

Using TTT-diagrams, one can describe the *crystallization of glasses* (completion of crystallization), as shown in Figure 9.20, or the *variation in resulting microstructure*, as indicated in the TTT-diagram given in Figure 9.21, which is related to the $Fe\text{-}Fe_3C$ phase diagram.

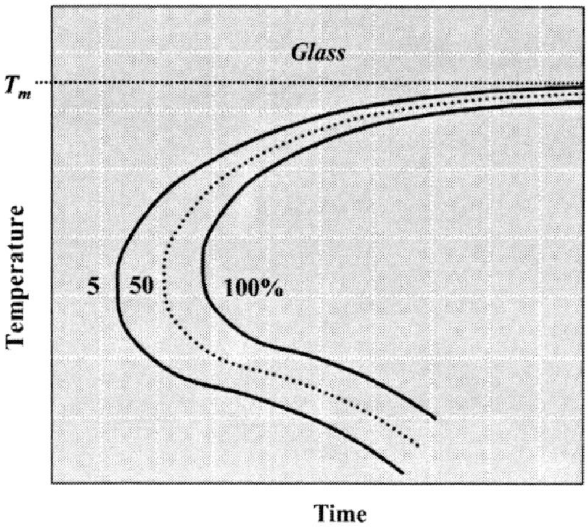

Fig. 9.20 TTT-diagram for the crystallization reaction of a glass (the different curves refer to various percent of completion of the reaction).

Note that the TTT-diagram in Figure 9.21 ($Fe\text{-}Fe_3C$ phase diagram) describes phase transformations that occur in the solid state.

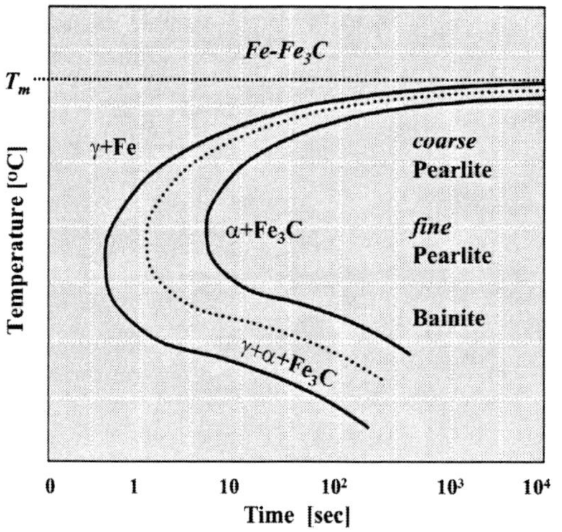

Fig. 9.21 TTT-diagram for the eutectoid $Fe\text{-}Fe_3C$ phase diagram, indicating the different microstructures that form depending on the annealing history (e.g., Bainite versus Pearlite).

Questions

What is the main driving force of sintering?

What are the mechanisms that contribute to shrinkage (densification) of ceramics?

How do you process the microstructure for transformation toughening of zirconia?

What is the mechanism of transformation toughening of zirconia?

What is the effect of Na_2O addition to pure SiO_2-glass?

How can compressive stresses be introduced in a glass surface?

How can you prevent crystallization of a silicate glass?

10 Polymers

The word **polymer** literally means "many mers" or "many parts". A common *synonym* for polymers is "plastics", a name that derives from the ease of formability of polymers. A *mer* (monomer) is the individual building unit of a polymer. Combining many such small organic molecules (mers) via chemical bonding leads to a polymeric solid.

Polymers represent an important group of engineering materials, which continuously increase their effect on our daily life. They are composed of *long chains* or *network molecules* and are usually based on **hydrocarbons** (oil; technology from the 20th century), where the smaller units (mers) such as C_2H_4 are linked together (Figure 10.1).

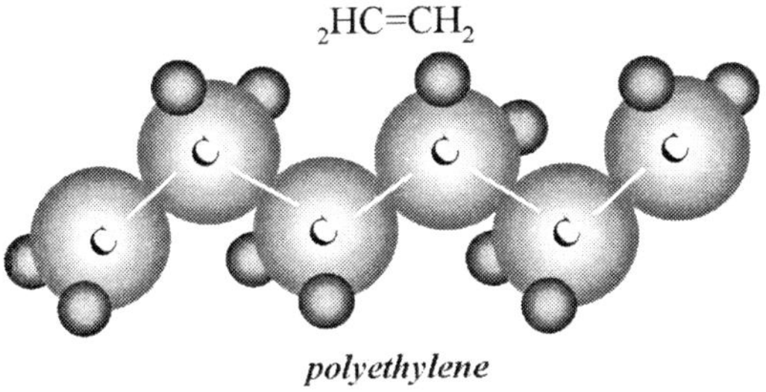

Fig. 10.1 Structural model of a small segment of a polyethylene chain $(C_2H_4)_n$; only three mers are linked here.

As shown in Figure 10.2, individual mers, $_2HC{=}CH_2$ (ethylene), are linked together forming long **chains** (polyethylene). The bonding within a single chain is predominantly **covalent**, while bonding between individual chains involves much weaker secondary **van der Waals** bonding.

10.1 Polymerization

The process of polymer synthesis is called **polymerization**. It is a chemical reaction in which individual molecules (monomers) are converted into large weight molecules (polymers), as schematically shown in Figure 10.2.

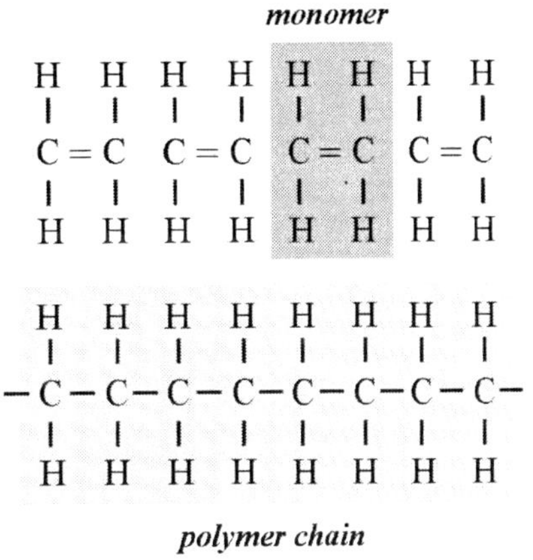

Fig. 10.2 Polymerization is the process where large molecules (chains; bottom) are produced from small building blocks (mers; top).

There are two major reaction schemes with quite different *kinetics*, commonly employed to initiate the polymerization reaction:

- **chain growth**, or *addition polymerization*, which involves rapid chain growth by the addition of free radicals
- **step growth**, or *condensation polymerization*, which is a much slower process and involves a reaction between different monomers.

Addition polymerization is a rather fast process promoting chain formation by the reaction between *chemically activated* monomers. In this reaction, an important feature of the monomer, which will be joined with similar small organic molecules, is the presence of a **reactive site.**

In *chain growth*, the critical feature of the monomer is the **carbon double bond**, while in *step growth* reactive **functional groups** are essential.

Polymerization of monomers to yield chain growth is obtained by "**opening up**" the C=C double bond within the monomer so it is available for the formation of long chains (*chain polymerization*).

Free radical polymerization is the most common method to create a reactive site and to open up the C=C double bond (see Figure 10.3). The reaction is initiated by adding a compound to the monomer solution that forms free radicals (for example hydrogen peroxide, H_2O_2). The free radicals attack the monomer and open up the C=C double bond.

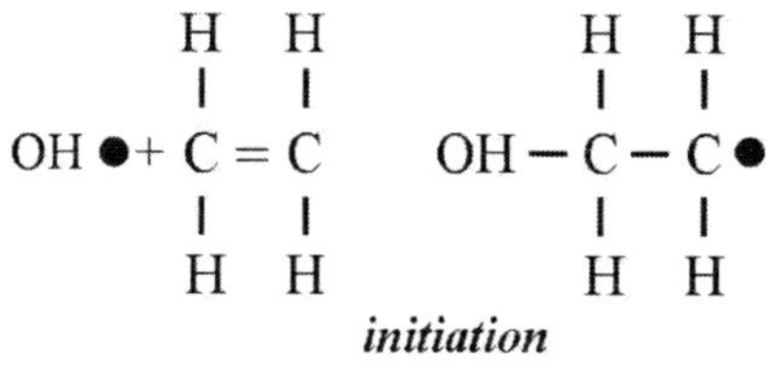

Fig. 10.3 Initiation of the polymerization reaction (chain growth) due to the addition of free radicals.

The addition of a free radical molecule, a molecule with an unpaired electron ($H_2O_2 \rightarrow 2OH\bullet$), initiates the polymerization reaction by creating a reactive site (unpaired electron) at the monomer molecule itself. This molecule then acts similar to the initially added free radical promoting the reaction. Propagation of chain growth occurs, since on the addition of each monomer unit, the unpaired electron moves to the end of the chain (Figure 10.4).

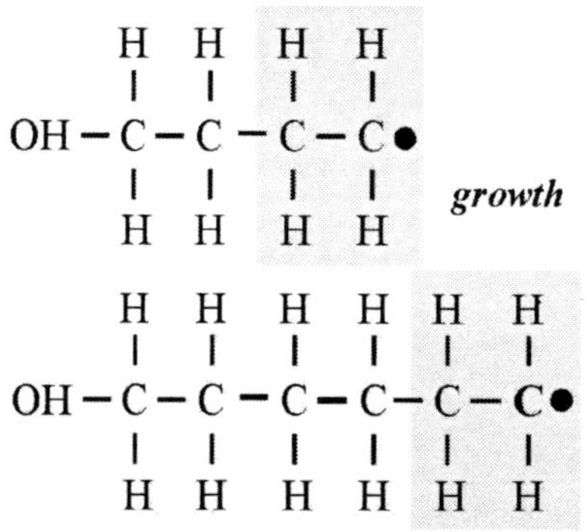

Fig. 10.4 Propagation of the polymerization reaction in which the unpaired electron moves to the end of the chain.

This reaction proceeds for quite some time and produces very large molecule chains with high molecular mass. However, the reaction is *terminated*, when a second free hydroxyl radical recombines with the unpaired electron of the growing monomer chain (Figure 10.5).

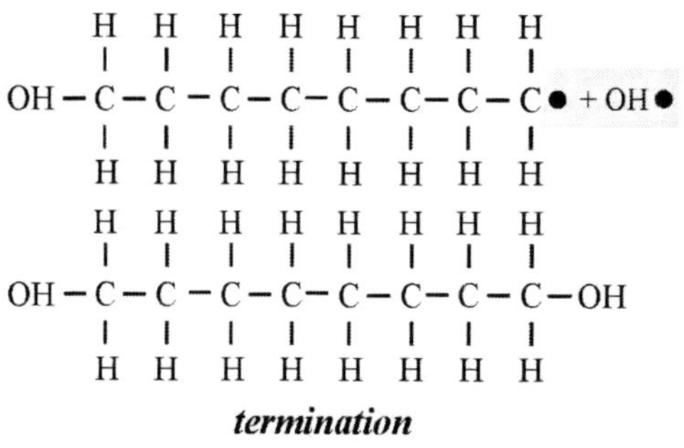

Fig. 10.5 Termination of the polymerization reaction due to recombination with another free hydroxyl radical.

Once monomers have been polymerized and long chains are produced, their chains are **jumbled together**, not unlike dreadlocks or spaghetti (Figure 10.6).

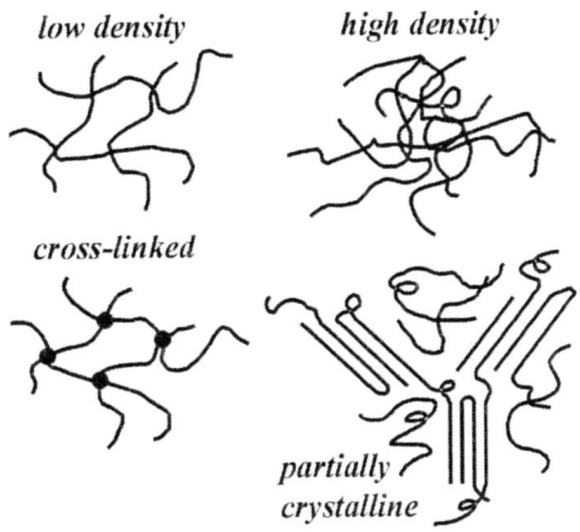

Fig. 10.6 Schematic of possible arrangements of long chains within the polymer.

Condensation polymerization, or step growth, is characterized by chemical reactions between different monomers; a rather slow process. Monomers, dimers, trimers etc. may all react together. In this case (Figures 10.7 and 10.8), two phenol molecules are linked by one formaldehyde molecule in a chemical reaction.

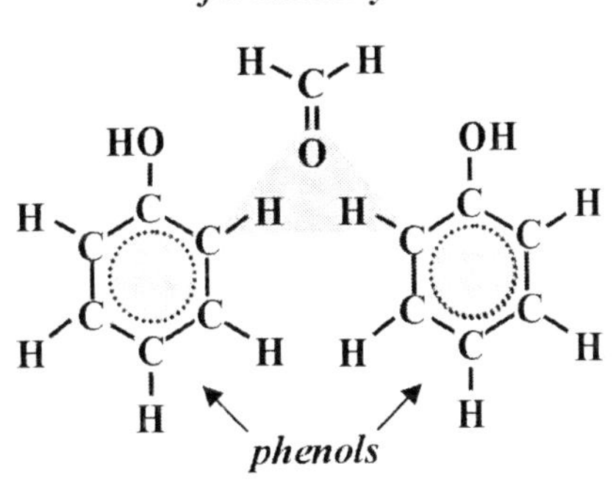

Fig. 10.7 Step growth reaction between two phenol molecules and one formaldehyde molecule.

Each phenol gives up a hydrogen atom and the formaldehyde gives up an oxygen atom. A water molecule is produced as a **by-product**. The common occurrence of by-products in these step growth reactions resulted in the term "condensation polymerization".

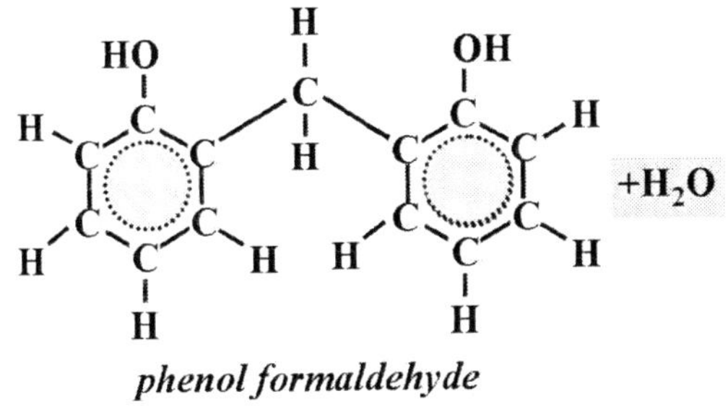

Fig. 10.8 Formation of phenol formaldehyde, Bakelite, via the step growth reaction (see also Figure 10.7).

The **functionality** of a polymer is defined as the number of active sites (or points of contact) of the mer. Functionality determines how a monomer polymerizes; for example, the ethylene mer is **bifunctional** leading to linear chain growth, while the phenol molecule has several potential active sites and thus is **polyfunctional**. The latter leads to a 3D-network molecular structure.

Note that a bifunctional monomer will produce a linear molecule by *either* chain or step growth, while a polyfunctional monomer will produce a network structure, **network polymerization**, by either of the two reactions.

In the example shown in Figures 10.7 and 10.8, two polyfunctional phenol monomers react with formaldehyde, resulting in the formation of *phenol formaldehyde*, **Bakelite**, which is a thermosetting phenolic plastic (see also following section).

10.2 Classification of Polymers

Polymeric materials can be classified into two major industrially important groups:

- **Elastomers** which have a huge reversible elastic deformation capability. Their mechanical behavior is analogous to natural rubber; they return to their original shape after stress is applied.
- **Plastics** that are a large and varied group of synthetic polymers, which are shaped by forming or molding. They reveal a far greater stiffness but lack the pronounced reversible elasticity.

Moreover, one distinguishes between *two types of plastics*, depending on how they are structurally bonded:

- **Thermoplastics** are typically linear polymers (including those that are branched but not cross-linked). They become soft and deformable upon heating and can easily be shaped into complex forms by melting or injection molding. ***Note*** that thermoplastics can be *reversibly reheated* and *reshaped* a number of times. One most common example is polyethylene.
- **Thermosetting plastics** (*thermosets*) are more rigid and are formed into a *permanent* shape and cured or "set" during a chemical reaction, which is **irreversible**. They cannot be remelted or reformed into another shape, but degrade or decompose when heated to high temperature. Thermosets are characterized by a rigid network-like structure and sometimes contain N, O, or S covalently bonded into the thermoset network. One example is the phenolic thermosetting plastic Bakelite.

Common ***thermoplastic polymers*** are polyethylene, polypropylene, polyvinyl chloride, polycarbonates, polystyrene, nylons, acetals, cellulosis, and tetrafluoroethylene (Teflon).

Common ***thermosetting plastics*** are polyurethane, phenolics, polyesters, amino resins, epoxies, isoprene (natural rubber), chloroprene (Neoprene), and silicones.

10.3 Degree of Polymerization

The degree of polymerization is the number of mers, n, which compose the polymeric molecules and is related to the chain length and the molecular weight. It can be determined from viscosity measurements or light scattering experiments.

The degree of polymerization represents an **average value** due to the statistical distribution of the molecular weight (chain length) within the polymeric solid. The number of n typically ranges between 2,500 and 35,000.

The average molecular weight, $\overline{M}_m$, of a polymer is the sum of the weight fractions times their mean molecular weight for each particular range divided by the sum of the weight fractions:

$$\overline{M}_m = \frac{\sum f_i M_i}{\sum f_i} \quad (10.1)$$

where M_i is the mean molecular weight for each molecular weight range and f_i is the weight fraction of the corresponding weight range (see also Table 10.1).

The average molecular mass of polymeric macromolecules is 100,000-700,000 g/mol (compare to metals with an atomic mass up to 300 g/mol).

Tab. 10.1 Average molecular weight of a polymeric solid.

Average Molecular Weight			
Weight Range	M_i	f_i	f_iM_i
5,000-10,000	**7,500**	**0.10**	**750**
10,000-15,000	**12,500**	**0.20**	**2,500**
15,000-20,000	**17,500**	**0.25**	**4,375**
20,000-25,000	**22,500**	**0.20**	**4,500**
25,000-30,000	**27,500**	**0.15**	**4,125**
30,000-35,000	**32,500**	**0.10**	**3,250**
		Σ=1.00	**Σ=19,500**

Example 10.1

20% are in the M_i range 10,000-15,000

30% are in the M_i range 15,000-20,000

50% are in the M_i range 20,000-25,000

it follows:

$$M_{20\%} = 12{,}500,\ f_{20\%} = 0.2$$
$$M_{30\%} = 17{,}500,\ f_{30\%} = 0.3$$
$$M_{50\%} = 22{,}500,\ f_{50\%} = 0.5$$

$$\overline{M}_m = \frac{0.2(12{,}500) + 0.3(17{,}500) + 0.5(22{,}500)}{1.00}$$

$$\overline{M}_m = 19{,}000\ g/mol\,.$$

10.4 Structural Features

There are many possibilities with respect to polymer chemistry; for example, hydrogen can be replaced with another atom or molecular group (R) such as Cl, CH_3, or a benzene ring, as shown in Figure 10.9.

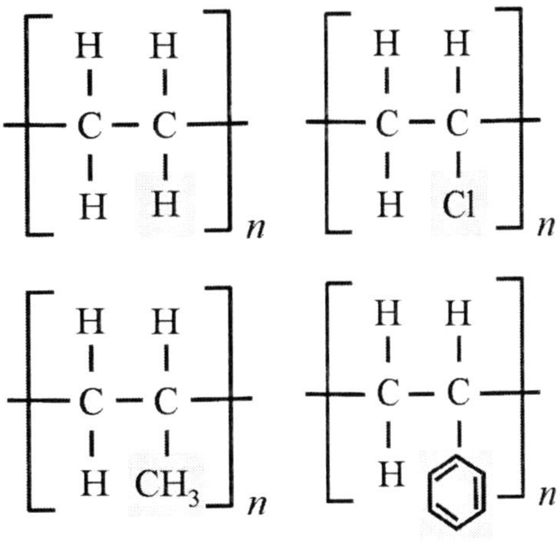

Fig. 10.9 Replacement of one hydrogen atom by other atoms or molecules leading to different polymers.

If the polymeric material is composed of $(C_2H_4)_n$ with R=H, a clear to whitish translucent thermoplastic, **polyethylene** (PE), is formed, which is used as containers (bottles), electrical insulation, electrical tubing, or house ware goods.

Replacing hydrogen by chlorine, R=Cl, produces an amorphous polymer, **polyvinyl chloride** (PVC), with reduced flexibility (steric hindrance due to the large chlorine ion). PVC can only be used for a few applications without the addition of other compounds (fillers, pigments).

Substituting a methyl group, R=CH_3, for hydrogen leads to a stronger but even less flexible polymer, **polypropylene** (PP). PP is used as packaging material, lab ware, auto interiors, or battery housings.

Incorporation of a benzene ring into the polymer backbone, R=C_6H_6, creates a clear, tasteless and odorless polymer, **polystyrene** (PS), which is relatively brittle. Typical applications include automobile interior parts, appliance housings, or dials and knobs.

Depending on the position of the side groups, R, one distinguishes three different **configurations** (see also Figure 10.10):

- **isotactic**; regular and all along one side
- **syndiotactic**; alternating along opposite sides
- **atactic**; irregularly placed.

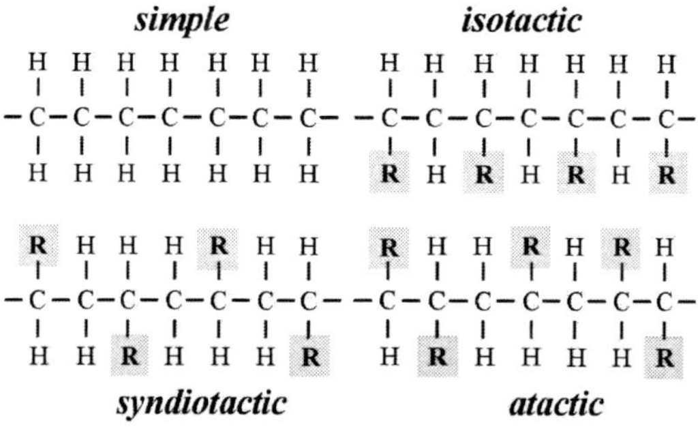

Fig. 10.10 Different polymer configurations depending on the position of the side group R.

When R, instead of being an atom or a molecule, represents another polymer chain, the first step of the transition from a *linear chain* structure to a *cross-linked* network structure is completed. This initial process is termed **branching** (Figure 10.11).

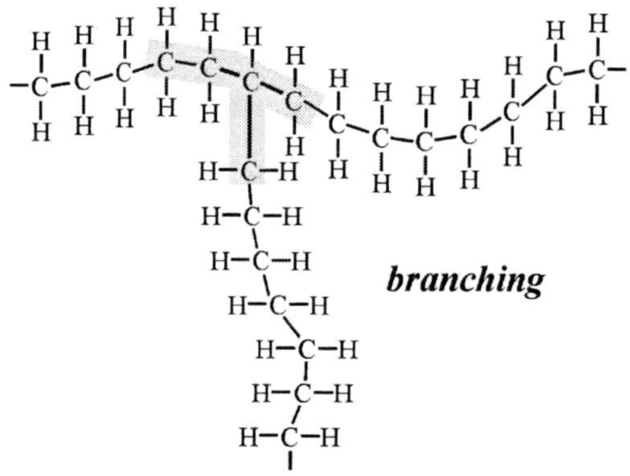

Fig. 10.11 First step in the transition from a linear to a network structure due to the process of branching.

The *complete transition* from a linear to a network structure is produced by **cross-linking** individual polymer chains. One important industrial example of the cross-linking process is **vulcanization** of isoprene (natural rubber; Goodyear 1839), shown in Figure 10.12.

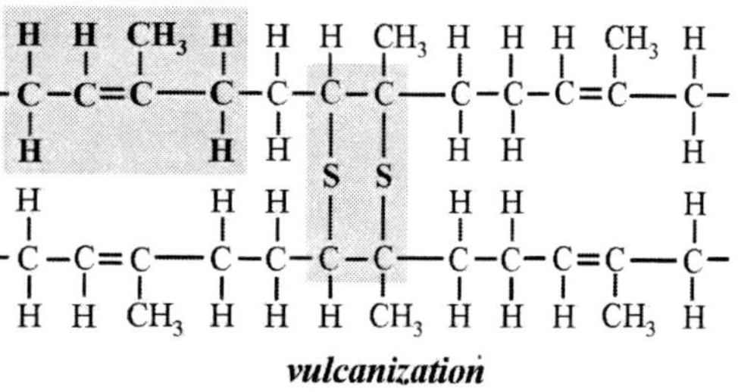

Fig. 10.12 Vulcanization (cross-linking) of isoprene chains via the addition of sulfur.

The *bifunctional* **isoprene** mer forms a linear chain structure; however, it still contains a C=C double bond after polymerization (Figure 10.12). The addition of sulfur atoms during processing opens up the C=C double bond (in two chains) and enables covalent bonding between two adjacent mers via two sulfur atoms.

In general, **polymer properties** change from soft to rigid (hard) with increasing degree of cross-linking. Strength and softening temperature of polymers increase with increasing degree of polymerization. Network structures are more rigid than chain molecules, since the latter can easily slide past one another. With increasing complexity of the polymer structure, crystallinity is commonly inhibited, resulting in amorphous structures.

Depending on how different monomer units ("blocks") are repeated within the chain, one further distinguishes between homo- and copolymers:

- **homopolymers** consist of a polymeric material with polymer chains composed of only a single repeating unit: AAAAAAAA.
- **copolymers** consist of polymer chains made up by two or more chemically different repeating units that can differ in sequence:
  - **alternating copolymer**: ABABABAB or ABBBABBBABBB
  - **random copolymer**: AABBBABAAB or ABBABABABBBBA
  - **block copolymer**: AAAAAABBBBBB or AAAABBBBBAAAA
  - **graft copolymer**: one long chain of polymer A with many appendages of polymer B: AAAAAAAAAAAAAAA
B B B
B B B
- **polymer blends** are another form of alloyed polymers in which different already formed polymer molecules (chains) are mixed together.

The main reason for **blending** different polymers (*alloying*, which is similar to metallic alloying; however, with limited solubility) is to *combine specific properties* of the polymers utilized.

One prominent example is the polymer blend **ABS** (Figure 10.13). Here, three pre-formed polymers are alloyed together: (i) acrylonitrile (A), (ii) butadiene (B), and (iii) styrene (C). The resulting material is a **graft copolymer** with the main chain made up by butadiene.
The component acrylonitrile gives the polymer chemical resistance and improved toughness, butadiene increases the impact strength of the polymer, while styrene enhances the rigidity of the final product.

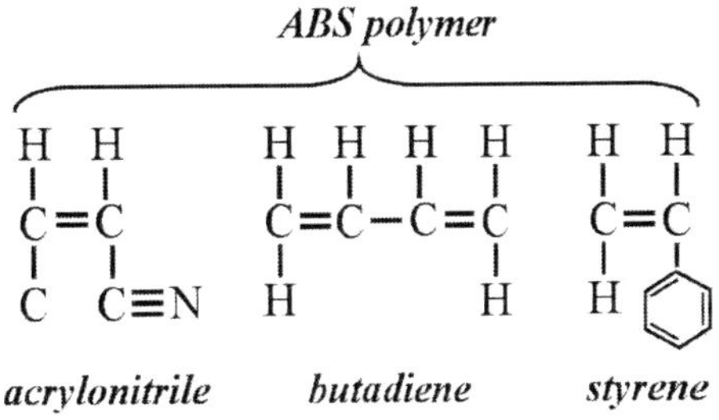

Fig. 10.13 Shown are the three monomer units composing the ABS thermoplastic polymer blend.

Applications of the ABS polymer blend are pipe and drain fittings, automotive parts, door liners for refrigerators, computer and telephone housings, electrical conduit, and luggage casings.

Crystallinity of polymers is predominantly relevant for *thermoplastic polymers*, since cross-linked structures are generally non-crystalline (high complexity of polymer structure). The crystallization behavior of polymers is related to the *amount of order* within the polymer. For example, **isotactic** and **syndiotactic** polymers tend to *crystallize*, while **atactic** polymers commonly remain *amorphous*.

Resulting mechanical properties of thermoplastic polymers are strongly determined by the amount of crystallization that occurred upon processing. Usually, an increase in crystallinity results in higher strength and higher stiffness. For polymers, just like for glasses, the ***glass transition temperature***, T_g, is used to define the transition in properties, for example, from rubbery to flexible. Hence, polymer crystallinity affects the flexibility of the polymer component.

Note that, unlike ceramics and metals, crystallization of thermoplastics can range from 0 to 100 % (the phenomenon of polymer crystallization is, however, not yet well understood).

10.5 Mechanical Properties

When external stress is applied to polymers below the transition temperature, T_g, elastic deformation is observed due to **uncoiling** of individual polymer chains. Applying stress above T_g results in plastic deformation via **sliding** of polymeric chains past one another and breaking secondary bonds between molecules (Figure 10.14).

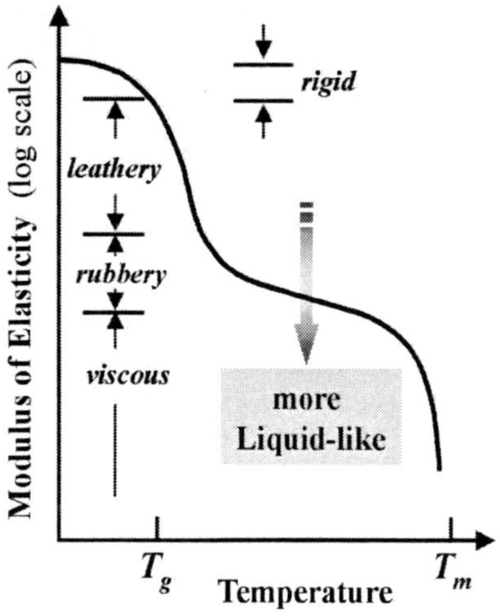

Fig. 10.14 Modulus of elasticity as a function of temperature of a thermoplastic polymer.

Due to a high fraction of secondary bonding, the elastic modulus of polymers is orders of magnitude lower than those of metals or ceramics. Figure 10.14 shows four distinct regions of modulus change versus temperature. At low temperatures (below T_g), a **rigid** modulus results in brittle behavior of the polymer (deformation is *linear elastic*).

In the range of T_g, the so-called **leathery** region, the modulus drops substantially. Qualitatively, this region can be interpreted as the onset of long-range molecular coordinated motion, since a larger number of molecules have sufficient thermal energy to move in a coordinated manner. The polymer can be extensively deformed in this temperature range and *slowly* returns to its original shape upon unloading.

Just above T_g, a plateau-like **rubbery** region is shown in Figure 10.14. Here, extensive deformation is possible with *rapid* spring back to the original shape when stress is removed. ***Note*** that in the leathery and rubbery region, extensive *non-linear deformation* of polymers can be fully recovered, which is by definition **elastic recovery**.

When the temperature approaches the melting point, T_m, the modulus of elasticity drops again. When the liquid-like **viscous** region is entered, deformation occurs via *viscous flow*. The overall mechanical response of polymers is hence **viscoelastic**.

Note that the **modulus of elasticity** can be greater (factor of 2), when measured dynamically. It increases with frequency since less sliding of chains occurs. The modulus of elasticity, E_{elast}, can be determined using the same specimen geometry used for modulus of rupture tests:

$$E_{elast} = \frac{L^3\, m}{4\, b\, h^3} \qquad (10.2)$$

where m is the slope of the tangent to the initial straight portion of the stress-strain curve, L is the distance between fixtures, b and h are specimen width and height, respectively.

Figures 10.15 and 10.16 show the effect of partial crystallinity and increased cross-linking on the modulus of elasticity.

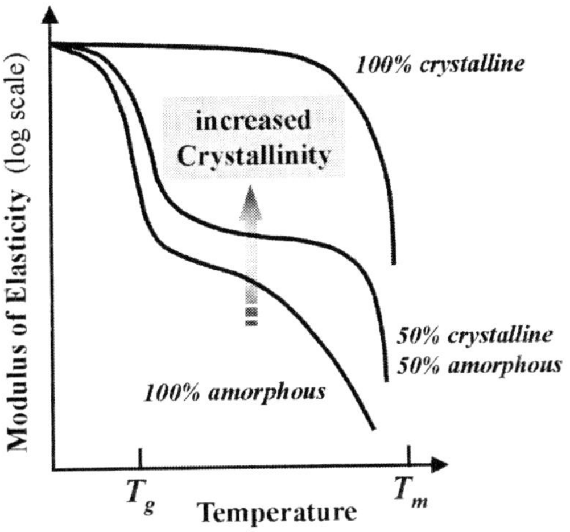

Fig. 10.15 Rigidity as a function of temperature for thermoplastic polymers with different crystallinity.

When the polymer is fully **crystalline** (Figure 10.15), there is no rubbery region. The material is initially rigid but softens when temperature approaches T_m, which is a similar behavior shown by crystalline metals and ceramics. A fully amorphous polymeric material, on the other hand, shows the general curve shape shown in Figure 10.14, with a pronounced leathery and rubbery region.

Increased **cross-linking** of the polymer network results in a similar mechanical behavior as an increase in crystallinity (compare Figures 10.16 and 10.15). This similarity is a consequence of the increase in rigidity of the polymer network structure (due to increasing crystallization or cross-linking). Cross-linked polymers do not experience chain uncoiling and sliding to the same degree as polymers with no cross-linking.

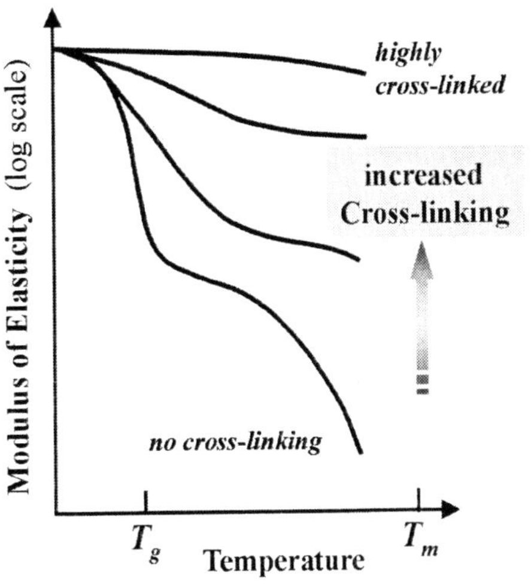

Fig. 10.16 Rigidity of thermoplastic polymers as a function of temperature with different degrees of cross-linking.

The modulus of elasticity, E_{elast}, versus temperature curve of **elastomers** shows a *huge rubbery region*, as illustrated in Figure 10.17, which resembles the normal room-temperature behavior of elastomers.

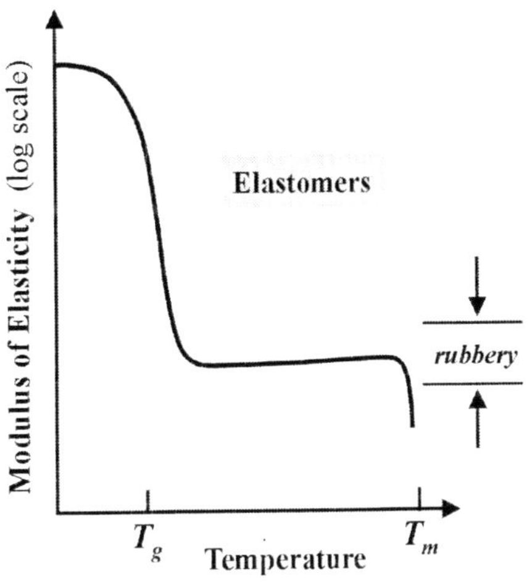

Fig. 10.17 Modulus of elasticity versus temperature of an elastomer. Note the pronounced rubbery region.

Note that the glass-transition temperature of elastomers is below room temperature. These materials (a subgroup of thermoplastic polymers) are an extreme example of a huge elastic deformation capability (up to 800%), due to uncoiling of long polymer chains. However, the corresponding *stress-strain response* is **non-linear**, as shown in Figure 10.18.

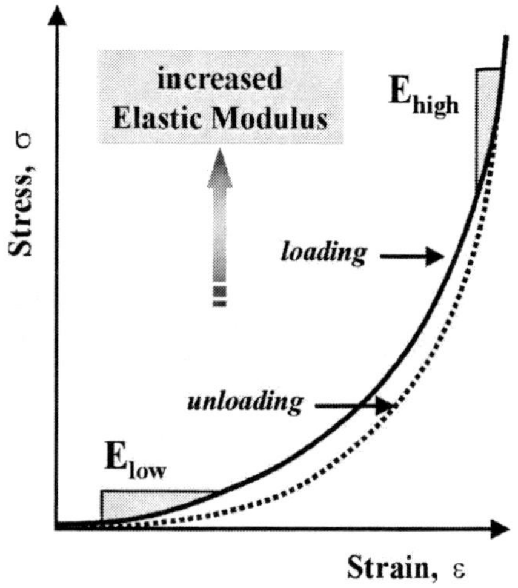

Fig. 10.18 Stress-strain curve of an elastomer showing non-linear elasticity.

For **low strains** (up to ~15%), the elastic modulus, E, corresponds to small forces required to overcome secondary bonding among linear chains (weak Van der Waals bonding) and to uncoil the large molecules. For **high strains**, the elastic modulus rises sharply, indicating that greater forces are needed to further uncoil the chains and to stretch primary bonds in the molecule backbone (covalent bonding). *Note* that in both regions (low and high strains), a significant fraction of secondary bonding is involved in the deformation mechanism, resulting in a *much lower elastic modulus* of polymers as compared to metals and ceramics.

It is important to emphasize that the deformation behavior shown in Figure 10.18 reflects **temporary** (elastic) deformation. When stress is removed, the polymer molecules **recoil** to their original length. The dashed line in Figure 10.18 shows that the recoiling of polymer chains has a slightly different path than uncoiling of the molecules during loading. Such a difference in loading and unloading behavior is termed **hysteresis**. If the stress level is further increased, secondary bonds break, which originates chain slippage (plastic deformation).

The process of uncoiling polymer chains is related to their **root-mean-square length,** $\bar{L}$, which is given by:

$$\bar{L} = l \cdot \sqrt{m} \qquad (10.3)$$

where l is the length of a single bond in the backbone of the polymer molecule and m is the number of bonds (see also Figure 10.19).

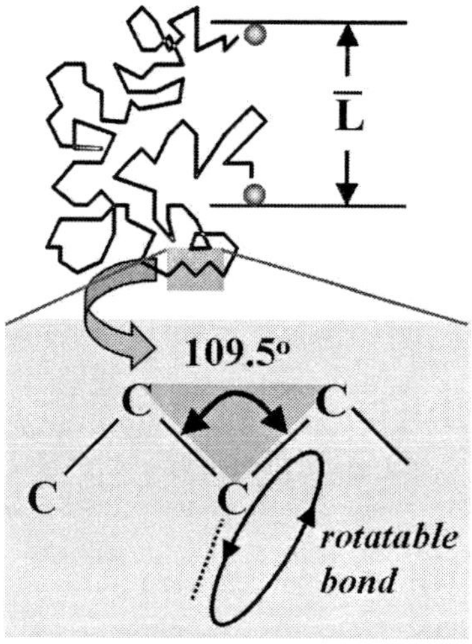

Fig. 10.19 Schematic showing the root-mean-square length, $\bar{L}$, of a kinked and coiled polymer chain.

As shown in Figure 10.19, each bond angle between three adjacent C-atoms is 109.5°, but this angle can be rotated freely, resulting in a kinked and coiled molecule configuration. $\overline{L}$ represents the effective length of the linear molecule, while the hypothetical length parameter **extended length** , L_{ext}, describes the length of a molecule that is extended as much as possible without distortion of bond angles:

$$L_{ext} = ml \cdot \sin\frac{109.5^0}{2} \qquad (10.4)$$

For typical bifunctional polymers such as PE or PVC, there are two bond lengths per mer, $m = 2n$, where n equals the degree of polymerization.
It is this **uncoiling process** (stretching of polymer chains) that is responsible for the huge elasticity monitored when deforming a rubber band for example.

10.5.1 Creep and Stress Relaxation

Creep deformation is defined as the increase in strain with time under constant load and constant temperature. In contrast, **stress relaxation** is the decrease of an applied load (stress) under constant strain and constant temperature.

Many polymer materials are prone to creep deformation (time-dependent deformation under constant externally applied load), even at low temperatures (RT); generally above T_g. Creep is an important designing factor when using polymers, given their relatively low melting (decomposition) temperatures. The mechanism of creep deformation is a *combination* of elastic and plastic deformation, which is termed **viscoelastic** deformation.

Stress relaxation is also an important design consideration for polymers. A familiar example is a rubber band that was stretched for some time and that does not spring back to its original shape when the stress is removed.

The **mechanism** of stress relaxation (decreased stress under constant strain) is **viscous flow**. The applied constant strain induces the polymer chains to slide past one another. This viscous flow converts part of the elastic strain into *non-recoverable* (permanent) plastic deformation. Stress relaxation is characterized by a **relaxation time**, τ, which is defined as the time necessary to reduce the initially applied stress, σ_0, to $0.37 = 1/e$ of its value. The exponential decay of the applied stress is given by:

$$\sigma = \sigma_0 \cdot e^{\frac{-t}{\tau}} \qquad (10.5)$$

where σ is the stress at time t, σ_0 is the initially applied stress at time $t = 0$ and τ is the relaxation time (a property of the material). ***Note*** that, similar to creep, stress relaxation is an Arrhenius-like phenomenon, as shown in equation 10.6:

$$\frac{1}{\tau} = C \cdot e^{-Q/RT} \qquad (10.6)$$

where C is a pre-exponential constant, R is the universal gas constant, T the absolute temperature (in [K]) and Q is the activation energy for viscous flow.

Example 10.2

The relaxation time, τ, for a rubber band at 25 °C is 60 days. The initial stress applied is 2 MPa; how many days will be required to relax the stress to 1 MPa ?

$$1\,MPa = 2\,MPa \cdot e^{-t/60\,days}$$
$$t = -(60\,days)(\ln 0.5) = 41.5\,days.$$

10.5.2 Fracture

Fracture mechanics for **rigid** and thus brittle polymers is applied similarly to what was discussed for brittle ceramics and glasses. The major difference with respect to polymers is that there is *no atomistically sharp* crack tip, but a so-called **craze region** (Figure 10.20).

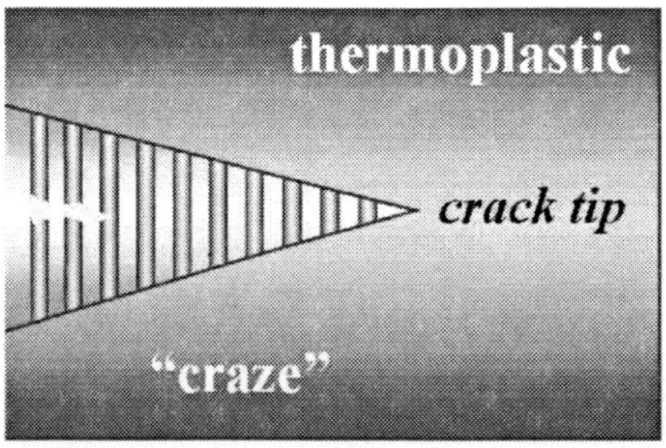

Fig. 10.20 Schematic illustrating the bridging of polymeric "fibers" when crack propagation occurs.

The craze region supports loading, because the fracture surfaces are **bridged** by several fine polymer fibers (stretched across the crack). The area involved by crazing can scatter light and, therefore, becomes white (*whitening process*). Due to this process, the *energy required* to drive (propagate) a crack is orders of magnitude *higher* than the enrgy necessary to break single C-C bonds.

Note that T_g determines the mechanical response of the polymer: below T_g polymers commonly reveal brittle fracture, while above T_g amorphous polymers exhibit large strains before fracturing. At intermediate temperatures, fracture occurs in a more ductile-like manner (necking).

10.6 Polymers in Brief

Processing of polymers in industry involves natural resource materials such as petroleum, natural gas, and coal.

Part-making of ***thermoplastics*** (soft and flexible upon heating) involves the following steps:

- **injection molding** (squeeze into mold)
- **extrusion** (exits through die)
- **thermoforming** (the mold clamps the hot polymer).

Part-making of ***thermosetting*** plastics (hard and rigid upon heating) involves:

- **compression molding** (heat and press)
- **transfer molding** (like compression molding, but here the material flows to another part of the die)
- **injection molding**.

Additives are used in polymeric technology to provide specific characteristic to the polymers:

- **plasticizers** are used to soften the polymer material
- **fillers** are commonly used for volume replacement and to improve polymer stability, but mainly to reduce costs (30% of automobile tires is filler)
- **reinforcements** enhance strength and stiffness (addition of glass fibers)
- **stabilizers** are utilized to reduce polymer degradation during oxidation, thermal cycling or exposure to UV light
- **colorants** (pigments) are added to gain the desired coloration of the polymer component.

10.6.1 Thermoplastics

About 60 % of all thermoplastics are either polyethylene (PE, high or low density), polyvinyl chloride (PVC), polypropylene (PP) or polystyrene (PS). They are cheap (costs < \$0.75/lb) compared with Teflon (polytetrafluoroethylene), which is close to \$6.00/lb.

Polyethylene (PE) compises close to 30% of all plastic sales in the US. It has good toughness at RT, good flexibility even at low temperatures (-60 °C), good corrosion resistance, and is rather cheap.
Applications include containers, electrical insulators, chemical tubing, housewares, blow-molded bottles, or pool covers.

Polyvinyl Chloride (PVC) is the 2nd most consumed plastic in the US.
Applications are basically twofold: (i) rigid compounds such as pipes, sidings, window frames, gutters, interior moldings and trims, and electrical conduits. Rubber resins are sometimes added to rigid PVC mainly to improve flow properties during processing. Upon cooling to RT, rubbery particles are present in the compound, which can absorb energy resulting in a much higher impact strength (shock absorbing pads).

(ii) **Plasticized PVC**, where plasticizers such as phthalate esters are added to the PVC, result in increased flexibility of the component. The plasticizers must be miscible and compatible with PVC. However, their addition usually results in lower tensile strengths.
Applications are furniture, auto upholstery, interior wall coverings, rainwear, shoes, luggage, shower curtains, many automobile applications such as auto top coverings, floor mats, interior and exterior trims, garden hoses and appliance components.

Polypropylene (PP) is the 3rd most used polymer by weight. Due to the addition of CH_3 in place of H, the rotation of chains is hindered and the flexibility of PP goes down as compared with polyethyelene. On the other hand, strength increases and the melting point also slightly increases due to the presence of methyl groups (165=>177 °C). Polypropylene can be subjected to ~120 °C without deformation.
Applications are packaging, lab ware, bottles, battery housings, fender liners, heater ducts, or as hinges (good long-term flexibility).

Polystyrene (PS) is the 4th most used thermoplastic by weight. Costs are ~\$0.60/lb and applications are automobile interior parts, appliance housings, dials, knobs, containers, or foams.

Acrylonitrile Butadiene Styrene (ABS) has good impact and mechanical strength (30-50 MPa is the tensile strength with 10-70% elongation). Costs are ~\$1.00/lb and applications are pipe and drain fittings, automotive parts, computer and telephone housings, door liners for refrigerators, electrical conduit, or luggage casings.

Polymethylmethacrylate (PMMA) is also called acrylics. It covers only about 1% of the market by weight. Plexiglass or Lucite are common tradenames. PMMA is transparent to visible light and has high impact resistance.
Applications are glazing for aircraft and boats, skylights and windows, exterior lightings, advertising signs, auto tailgate lenses, or protective goggles.

Polytetrafluoroethylene (PTFE), or Teflon, which is a Dupont tradename. PTFE softens at 370 °C. Since the F atom is rather small, Teflon is very dense; 2.2 g/cm^3.

Applications of PTFE are chemically resistant pipe and pump parts, gaskets, seals, high-temperature cable insulation, molded electrical components, tape, or non-stick coatings.

Engineering Thermoplastics: there are three major high-performance engineering thermoplastics:

- thermoplastic polyesters (PBT, PET)
- polyamides (nylons)
- polycarbonates.

The two main thermoplastic polyesters are: (i) polybutylene terephthalate (PBT) and (ii) polyethylene terephthalate (PET).

PBT is used in a number of industrial applications such as pump impellers, housings and support brackets, irrigation valves and bodies, water meter chambers and components, high-energy ignition caps and rotors in automobiles, speedometer frames and gears. There are also a number of electrical applications: switches, relays, connectors, TV tuner components, high-voltage components, terminal boards, and integrated circuit boards.

PET is used as Dacron (fiber) or Mylar (film); the latter is the transparent foil that is also used for food packaging. PET fibers are used for clothing, carpeting, tire cords, magnetic tape, fibers, and films.

Polyamides (nylons) reveal high strength (approaching 100 MPa) when highly crystalline, due to good secondary bonding between the molecular chains.
Applications are unlubricated gears (for example in speedometer or windshield wiper gears), bearings, ropes, fabrics, containers for corrosive fluids (brake and power steering fluid reservoirs), and high-impact applications (engine fan blades).

Polycarbonates (Lexan and Merlon are tradenames) show extremely high impact strength. The incorporation of two methyl groups causes local 'steric hinderence' and repulsion, which results in high stiffness.
Applications are safety shields, machine parts (cams and gears), helmets, propellers, and hand-held power tool housings.

10.6.2 Thermosetting Polymers

Thermosetting polymers cannot be reheated and reformed, but generally have higher thermal stability as compared with thermoplastics. They also show higher rigidity and resistance to loading. Most thermosetting polymers contain a filler ingredient (wood flour, mica, glass or cellulose). The most important representative of this group is the phenolic thermosetting plastic, Bakelite (tradename). Applications are electrical equipment (switch gear, relay systems), or automotive parts such as power-assist brake components and transmission parts.

Epoxy resins barely shrink when they are cured, because they do not give off any by-products during curing. They show good adhesion to other materials, good chemical and environmental durability, good mechanical properties, and good electrical insulating properties.
Applications are protective coatings, insulators for high-voltage applications, adhesives, and the matrix for a number of composite materials.

Polyesters are commonly first processed as an unsaturated polyester, which still contains C=C covalent double bonds that can be used for cross-linking to form thermosetting composite materials. In combination with, for example, glass fibers, unsaturated polyester can be cross-linked to form a fiber-reinforced composite material of high strength (fiberglass composite).

Elastomers, or rubbers, are a special sub-group of thermosetting polymers, since they also cannot be reshaped via heat treatment. More than 50% of all elastomers are made out of natural rubber, polyisoprene. Natural rubber is a milky liquid known as latex, which is a suspension containing very small rubber particles. This suspension is then processed (dried and shaped) and then vulcanized, a process which was first performed by Goodyear in 1839.

Questions

What is the most common process to initiate addition polymerization?

What is the main difference between thermoplastics and thermosetting plastics?

How can individual isoprene chains be cross-linked?

Why do elastomers show a non-linearity in elasticity?

What is the mechanism of stress relaxation of polymers?

10.6.3 Mechanical Properties

Here, the mechanical properties of some important polymeric materials are compared to one representative metal and one ceramic.

Tensile Strength

1040 Carbon Steel	750 MPa
Al_2O_3	210
Polystyrene	48
Polyethylene LD	14
HD	28
PVC	41
ABS	28-48
Phenolics	52
Polyamids	95

Elastic Modulus

Al_2O_3	320000 MPa
Cu	120000
PS	3000
PE-LD	170
PE-HD	830
PVC	2800
Rubber	2
ABS	2400
Phenolics	6900

Poisson's ratio

Al_2O_3	0.26
Cu	0.33
Plastics	0.20-0.40
Elastomers	0.49-0.499

In general, the strength of polymeric materials strongly depends on: (i) molecular mass, (ii) degree of crystallinity, (iii) additional atoms or molecules incorporated into the structure (vinyl polymers), (iv) introduction of N, O, or S in place of C into the main chain (vulcanization) and (v) the addition of fillers (20-40 wt% of glass fibers is typical).

11 Composites

A composite material is defined as a **mixture** of two or more constituents (one phase or several phases). The constituents are not soluble in one another (a solid solution forming a metal alloy is not a composite). Composites can be described as materials:

- that consist of two or more *physically* and/or *chemically distinct*, suitably arranged or distributed phases, with an **interface** separating them
- with *characteristic properties* that are not possessed by any of the components in isolation.

Composites are a relatively new class of materials that combine two or more separate components. However, the concept of composite materials is ancient. The combination of two different materials to produce a new one with performance unattainable by the individual constituents was employed when adding straw to mud for building stronger **mud walls**. Some more recent examples of composite materials (before engineered materials became prominent) are:

- **carbon black** in **rubber**
- **steel rods** in **concrete**
- **cement/asphalt** mixed with **sand**
- **glass fibers** in **resin**.

While each component retains its identity, the new composite material displays macroscopic properties **superior** to its parent constituents, particularly in terms of mechanical properties (strength, stiffness) and hence in economic value.

Composite materials have been widely used by the **aerospace** and **defense** industries because of their superior performance:

- **strong** and **rigid**
- extremely **lightweight**
- **corrosion** resistant
- **environmental damage** resistant
- **fatigue** resistant
- **flexible** in **design** (capable of being formed into large and easy-to-integrate parts; Figure 11.1).

Fig. 11.1 Car body made of a fiber-reinforced polymer composite.

Certain properties can also be *tailored* into composites such as:

- increased **wear** resistance
- improved **acoustical**, **electrical,** and **thermal** performance
- **aesthetic** characteristics.

For all those reasons, composite materials show tremendous promise for structural applications in automobiles, appliances, sporting goods, agriculture, bridges, roads, and building materials.

One important example, for those living in Colorado, is skis and snowboards. They must be strong und flexible, yet stiff in particular directions, and should also show high impact resistance.

There are also numerous examples of composite materials found in **nature**: coconut palm leaves, bamboo, cellulose fibers in a lignin matrix (wood), or collagen fibers in a hydroxylapatite matrix (bone).

11.1 Fiber-Reinforced Composites

The **reinforcement** system in a composite strongly determines the strengthening mechanism. It is, therefore, convenient to classify composites according to the characteristics of their reinforcement, such as length, orientation, morphology, etc.

Here, however, we will only distinguish between **two** major reinforcements:

- **fiber-reinforced** composites
- **particle-reinforced** composites,

which have two important industrial representatives:

- **Fiberglass**
- **Concrete**.

11.1.1 Glass Fibers

Reinforcements are not necessarily used in the form of *long fibers*. They can be added to the matrix as **particles**, **whiskers**, **discontinuous fibers**, **sheets**, etc.

Note that a great majority of materials are, however, much *stronger* and *stiffer* in the fibrous form than in any other form. This explains the emphasis on using fibers in composite materials design.

Typical fiber properties are:

- low density
- very high strength (glass fiber = 2400 MPa, carbon fiber = 3100 MPa, Kevlar fiber = 3600 MPa)
- high stiffness
- flexible (to allow a variety of methods for processing)
- high aspect ratio (length/diameter), in order to transfer a large fraction of the applied load onto the fiber.

Note that fibers are added to a **ductile** matrix (polymers and metals) usually to improve *strength* and *stiffness*, while fibers are added to a **brittle** matrix (ceramics) to increase *fracture toughness*.

The most common and inexpensive fiber used in composite manufacturing is the **glass fiber**, usually used for reinforcement of polymer matrices.

Each **glass composition** development has led to an *optimal suitability* for specific applications, as for example the boron-silicate glass, **E-glass** ("electrical"). Its *low sodium* content is responsible for its especially **low electrical conductivity** and, hence, its application as a **dielectric** material. In addition, its *high boron* content makes it **chemically durable**. Therefore, E-glass is the most generally used fiber-glass composition (see also Table 11.1).

Typical **glass fiber** properties are:

- low density (~2.55 g/cc)
- reasonable tensile strength (~1.8 GPa)
- low stiffness (70 GPa) (strains more easily, therefore, high strength/density but low stiffness/density ratio
- good dimensional stability
- resistant to heat, cold and corrosion.

Tab. 11.1 Typical chemical composition (wt% of oxide addition) of four common glasses.

Glass Composition	SiO_2	Al_2O_3	CaO	MgO	Na_2O	B_2O_3
W	72	1	10	4	13	-
C	73	2	8	3	14	-
E	54	15	22	-	1	8
B	76	4	1	-	6	13

W=window glass, C=container glass, E=E-glass (fiber), B=borosilicate glass.

Note that glass fibers are susceptible to **static fatigue**, which means that they cannot withstand static loads for long periods of time. Moreover, moisture can dramatically decrease glass fiber strength.

11.1.2 Isostress and Isostrain

Glass fibers are available in three common *configurations*: (i) continuous fibers, (ii) discrete (chopped) fibers, or (iii) woven fabric (layered, laminate). Therefore, *composite properties* strongly depend on the **orientation** of the fibers within the component. As a result, the respective composite properties are highly **anisotropic**, as will be shown in the following for the elastic modulus (Young's modulus) of a fiber-reinforced polymer.

Under **isostrain** conditions, the externally applied load is **parallel** to the fibers, as shown in Figure 11.2.

$$\varepsilon_{comp} = \frac{\Delta l}{l_0} = \varepsilon_{fiber} = \varepsilon_{matrix} \qquad (11.1)$$

The load applied to the composite. P_c, equals the sum of the loads on the fiber, P_f and the matrix, P_m:

$$P_c = P_f + P_m \qquad P_c = \sigma_c A_c \qquad (11.2)$$

$$\sigma_c A_c = \sigma_f A_f + \sigma_m A_m \qquad (11.3)$$

using $$A_c = A_f + A_m \qquad (11.4)$$

and $$\varepsilon = \frac{\sigma}{E} \quad \text{or} \quad \sigma = \varepsilon E \qquad (11.5)$$

leads to $$\sigma_c A_c = \sigma_f A_f + \sigma_m A_m \qquad (11.6)$$

$$\varepsilon_c E_c A_c = \varepsilon_f E_f A_f + \varepsilon_m E_m A_m \qquad (11.7)$$

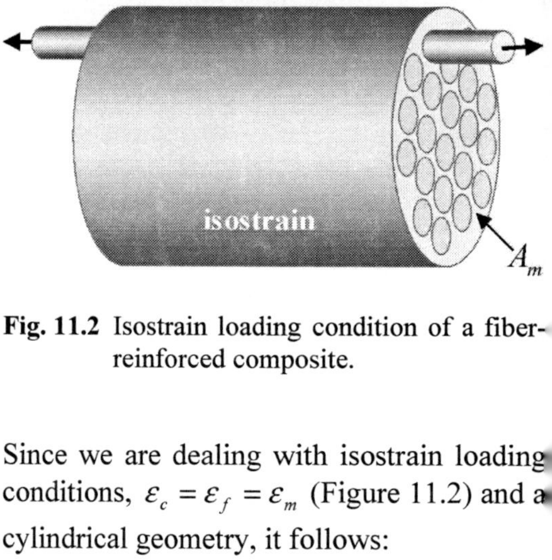

Fig. 11.2 Isostrain loading condition of a fiber-reinforced composite.

Since we are dealing with isostrain loading conditions, $\varepsilon_c = \varepsilon_f = \varepsilon_m$ (Figure 11.2) and a cylindrical geometry, it follows:

$$E_c = E_f \frac{A_f}{A_c} + E_m \frac{A_m}{A_c} \qquad (11.8)$$

and with the assumptions that:

$$\frac{A_f}{A_c} = \frac{v_f}{v_c} = v_f, \quad \frac{A_m}{A_c} = \frac{v_m}{v_c} = v_m \qquad (11.9)$$

this leads to the so-called **rule of mixtures** (ROM):

$$E_c = E_f v_f + E_m v_m \qquad (11.10)$$

The elastic modulus of the composite equals the sums of the volume fraction times the corresponding modulus of the fiber and the matrix and thus can easily be calculated.

The same **averaging**, as shown for the Young's modulus (equation 11.10), also applies for other properties such as:

- **diffusivity**
- **thermal conductivity**
- **electrical conductivity**.

Note that *interfacial bonding* between fiber and matrix is also a critical parameter, and will be addressed later when toughening of ceramic materials via the incorporation of long fibers is introduced.

It is interesting to address the question: which volume fraction of the total load applied to the composite, P_c, is carried by the uniaxially loaded fibers, P_f. The load fraction is given by:

$$\frac{P_f}{P_c} = \frac{\sigma_f A_f}{\sigma_c A_c} = \frac{\varepsilon_f E_f A_f}{\varepsilon_c E_c A_c} \qquad (11.11)$$

and with $\varepsilon_c = \varepsilon_f = \varepsilon_m$ and $\dfrac{A_f}{A_c} = \dfrac{v_f}{v_c} = v_f$

it follows: $$\frac{P_f}{P_c} = v_f \frac{E_f}{E_c} \qquad (11.12)$$

For the example given in Figure 11.3 (E-glass fiber-reinforced polymer (epoxy) composite), with an elastic modulus of the fiber of 72.4 GPa and of the composite of 52.8 GPa (70% vol% of fibers), it follows: $\mathbf{P_f/P_c = 0.96}$.

Hence, most of the externally applied load is transferred to the fibers, namely 96%.

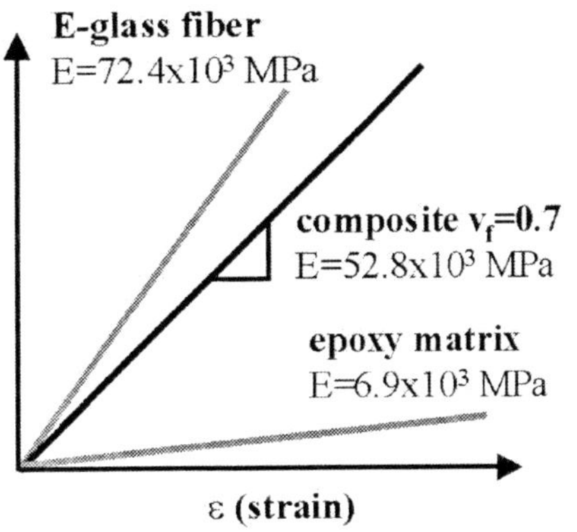

Fig. 11.3 Isostrain loading condition of a fiber-reinforced composite.

Now we assume that the external load is applied **perpendicular** to the fibers, as shown in Figure 11.4, which is termed **isostress** condition. ***Note*** that the fibers are schematically drawn as thin slabs embedded in the matrix.

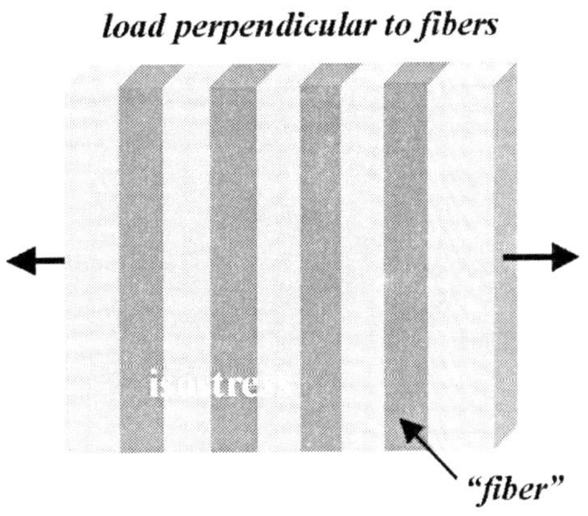

Fig. 11.4 Isostress loading condition of a fiber-reinforced composite.

If the external load is applied perpendicular to the fiber direction, then the stress that acts on the fibers equals the stress acting on the matrix, which leads to:

$$\sigma_c = \sigma_f = \sigma_m \quad (11.13)$$

$$\varepsilon_c E_c = \varepsilon_f E_f = \varepsilon_m E_m \quad (11.14)$$

The overall loading condition can be expressed in terms of total elongation:

$$\Delta L_c = \Delta L_f + \Delta L_m \quad (11.15)$$

$$L_m = A_m L_c \quad (11.16)$$

$$\frac{\Delta L_c}{L_c} = \frac{A_f \Delta L_f}{L_f} + \frac{A_m \Delta L_m}{L_m} \quad (11.17)$$

using $\varepsilon_c = \frac{\Delta L_c}{L_c}$ and assuming that the area equals the volume leads to:

$$\varepsilon_c = v_f \varepsilon_f + v_m \varepsilon_m \quad (11.18)$$

and with $\varepsilon_c = \frac{\sigma_c}{E_c}$ we derive:

$$\frac{\sigma_c}{E_c} = v_f \frac{\sigma_f}{E_f} + v_m \frac{\sigma_m}{E_m} \quad (11.19)$$

Since we are dealing with isostress loading conditions, $\sigma_c = \sigma_f = \sigma_m$, it follows:

$$\frac{1}{E_c} = \frac{v_f}{E_f} + \frac{v_m}{E_m} \quad (11.20)$$

and

$$E_c = \frac{E_f E_m}{v_f E_m + v_m E_f} \quad (11.21)$$

Here, the elastic modulus of the composite component equals the elastic modulus of the fiber, E_f, times that of the matrix, E_m, divided by the sum of the matrix modulus times volume fraction of the fiber and the fiber modulus times the volume fraction of matrix.

Comparing equation 11.21 with equation 11.10, a very different behavior of the elastic modulus of the composite depending on the volume fraction of fibers is found, as illustrated in Figure 11.5.

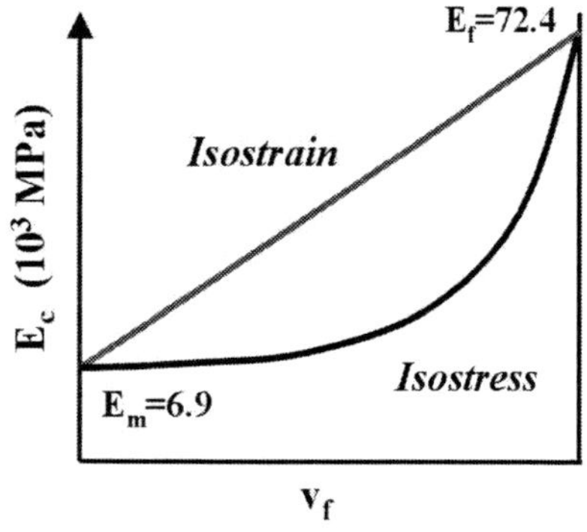

Fig. 11.5 Comparison of elastic modulus of the composite versus fiber volume fraction for isostrain and isostress loading conditions.

Under isostress loading conditions, an effective load transfer onto the fibers occurs only at rather high volume fractions of fibers. Hence, it is important for the engineer to know about fiber orientation within the composite and about the stress state the component will experience.

In case **aggregates** (particles) are used as reinforcement (Figure 11.6), the dependence of composite modulus, E_c, on the volume fraction of a reinforcement with high modulus, v_a, lies generally between the two extremes of **isostrain** and **isostress**, as shown in Figure 11.7.

The following equation 11.22 gives the general expression for the elastic modulus of a composite material with $n = +1$ for pure isostrain and $n = -1$ for pure isostress loading configuration:

$$E_c^{\,n} = v_f E_f^{\,n} + v_m E_m^{\,n} \quad (11.22)$$

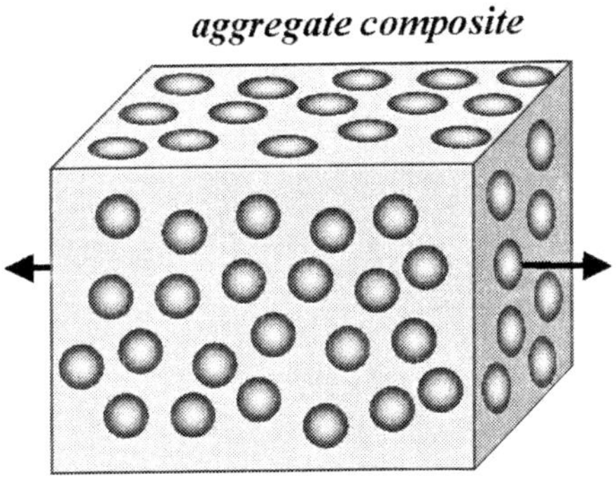

Fig. 11.6 Schematic of an aggregate composite where spherical particles are used as reinforcement.

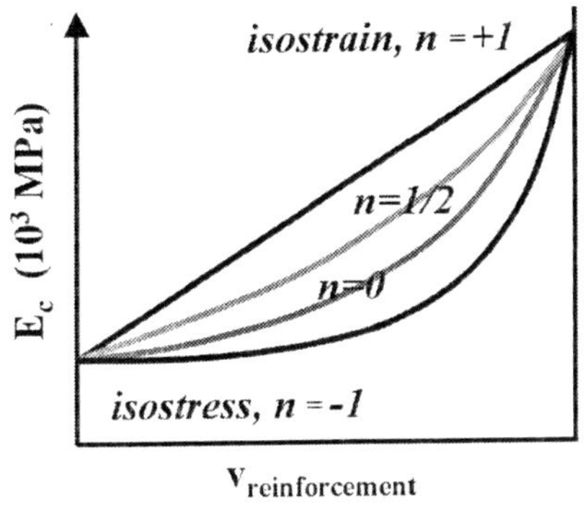

Fig. 11.7 Elastic modulus of a composite versus volume fraction of reinforcement. The modulus values of aggregate composites fall between the two curves of isotrain and isostress.

Example 11.1

Calculate the Young's modulus of a fiber-reinforced composite loaded under isostress conditions. The matrix is composed of 60 vol% Al_2O_3 (E = 375 GPa) and the fiber content is 40 vol% Al (E = 80 GPa); see also equation 11.21.

$$E_c = \frac{80 \cdot 375 \ GPa^2}{((0.4 \cdot 375) + (0.6 \cdot 80)) \ GPa}$$

$$E_c = \frac{30{,}000 \ GPa^2}{(150 + 48) \ GPa} = 151.51 \ GPa$$

11.1.3 Ceramic Fibers

Apart from most commonly used glass fibers, discussed in section 11.1.1, special application needs lead to the utilization of other fiber materials such as metal fibers, polymer fibers, or **ceramic fibers**. Carbon fiber composites find wide applications in the *aerospace* and *sporting goods* industries, due to their low densities (graphite ~2.3 g/cc). Furthermore, owing to the ability to tailor similar stiffness and strength properties to human bone, carbon fiber composites show specific advantages over metallic parts in **biomedical** applications (to a large extent in *orthopedics*).

For fiber production, the *graphitic structure* is preferred over the diamond polymorph. Therefore, carbon fibers are commonly made from **polyacrylonitrile** (PAN) or **Rayon** polymer precursors (Figure 11.8) by *carbonization* of precursor fibers, followed by *graphitization* at high temperature.

poly-acrylonitrile

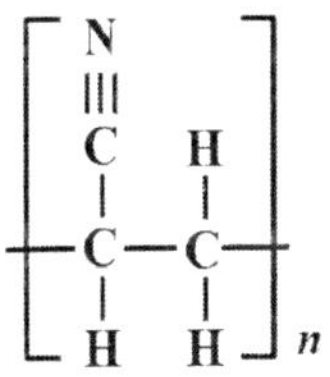

Rayon

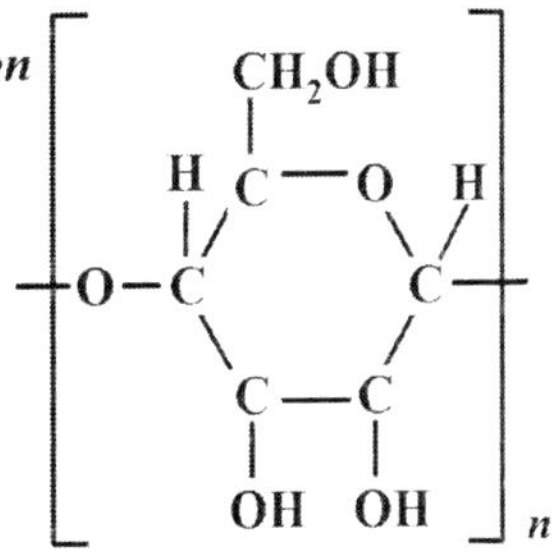

Fig. 11.8 Starting polymer precursors PAN and Rayon for carbon fiber processing.

Since the graphite structure is made of densely packed hexagonal layers stacked in a lamellar style, the *mechanical* and *thermal properties* are **anisotropic**. Therefore, controlling the orientation of the crystalline layers within the fibers is crucial.

The **Young's modulus** in the layer plane can be as high as 1000 GPa, while E along the c-axis is equal to about 35 GPa. Similarly, the **coefficient of thermal expansion** (CTE), α, of carbon fibers is highly anisotropic with:

- $\alpha_{traverse} = +5.5$ to $+8.4\ 10^{-6}$ K
- $\alpha_{parallel} = -0.5$ to $-1.3\ 10^{-6}$ K.

Ceramic fibers, such as alumina, Al_2O_3, or silicon carbide, SiC, are advantageous in very **high-temperature applications**, and also where environmental attack (corrosion) is an issue. ***Note*** that since ceramics have poor properties in *tension* and *shear*, many applications as reinforcement are in the particulate form (aggregate composite; see Figure 11.9).

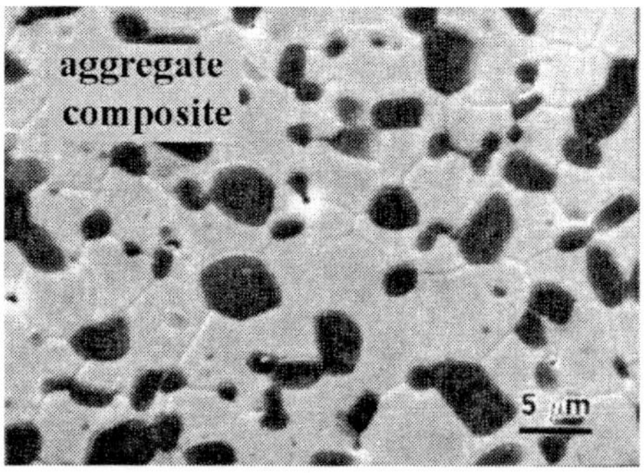

Fig. 11.9 A typical example of an aggregate composite with ZrO_2 particles (dark) dispersed in a Al_2O_3 matrix.

Moreover, processing of ceramic fibers in a reproducible way (high reliability) is very difficult and in most cases also very cost intensive.

11.1.4 Polymer Fibers

Since polymers are strongly bonded within the polymer chains (backbone), polymer fibers also show very **high strength** values when the high molecular weight chains are aligned *along the fiber axis*. Two examples of such high-strength fibers are UHMPE (ultra-high molecular weight polyethylene) called Spectra (made by Allied Corp.) and Kevlar (made by DuPont).

Spectra is a very light fiber (density ~0.97 g/cc) made from solutions and gels. Its primary disadvantage is the low melting point (~150 °C); but this is not an issue in biomedical applications. It has a stiffness of about 200 GPa. ***Note***, however, that most polymers have a much lower modulus of elasticity (see previous chapter).

Kevlar is an aramid (aromatic polyamide) composed of oriented aromatic chains, which makes it a rigid rod-like polymer (Figure 11.10). Kevlar has a very high T_g and, although very strong in tension, it has rather poor compression properties. Its stiffness can be as high as 125 GPa. Kevlar fibers are mostly used to **increase toughness** in otherwise brittle matrices.

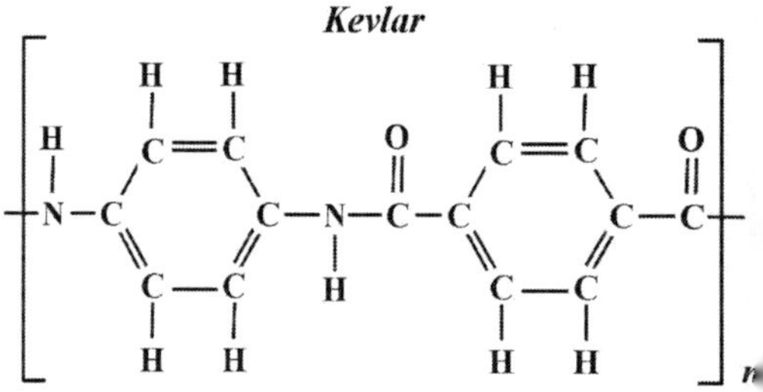

Fig. 11.10 Molecular structure of the polymer fiber, Kevlar.

11.1.5 Metal Fibers

Metal fibers, such as steel (rebar in concrete), tungsten (Figure 11.11), or boron have high strengths and show *very consistent* properties, unlike ceramic fibers. Since the density is very high for these fibers, they are not used in applications where weight needs to be reduced. Drawing very thin metallic fibers (less than 100 micron in diameter) is also very expensive, which strongly limits their application.

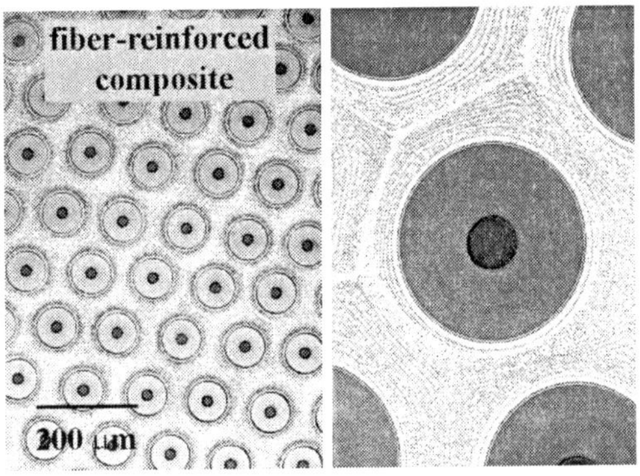

Fig. 11.11 Tungsten fibers (with a carbon core) embedded in a metal alloy matrix.

11.2 Composite Materials

A very large number of composite materials are known to date and the number of synthetic composites for very specific applications is continuously growing (favorable properties at low as well as high temperatures). Apart from man-made composites, evolution has also produced a vast number of *natural composites* such as bamboo, palm leaves, wood and, most important for *homo sapiens*, bone. Here, we will discuss three important composite materials: wood and bone, as representatives of the group of natural composites, and concrete, a widely used engineering composite (construction).

11.2.1 Wood (natural composite)

Wood is a naturally occurring fiber-reinforced composite, which serves as an excellent structural material. The weight of wood used in construction in the US *exceeds* the *combined total* of steel and concrete. There are *two categories* of wood: **soft-woods** and **hardwoods** (Table 11.2).

Tab. 11.2 Examples of the two categories of wood (softwoods and hardwoods).

Softwoods	Hardwoods
Cedar	Ash
Douglas fir	Birch
Hemlock	Hickory
Pine	Marple
Redwood	Oak
Spruce	Cherry

The fundamental difference between the two categories is their **seasonal nature**: *softwoods* are "evergreens" with needle-like leaves and exposed seeds, while *hardwoods* are deciduous in nature, since they lose their leaves annually and have covered seeds (nuts).

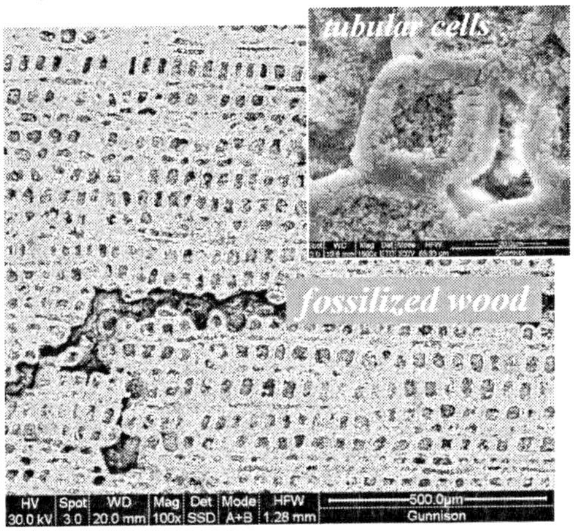

Fig. 11.12 SEM BSE-image of a fossilized softwood revealing the vertical tube-like cell structure (see also SE-image in inset).

For example, the **microstructure** of a pine tree, an important softwood, shows a large number of *tube-like cells*. These longitudinal cells are aligned with the vertical axis of the tree. One example of such *vertically aligned* cells is show in Figure 11.12, where the microstructure (cell structure) of a fossilized softwood from Florissant, CO is shown, cut perpendicular to the growth axis.

Note that it was Robert Hooke, who coined the term "*cell*" for these biological building blocks. There are also *radial cells* perpendicular to the longitudinal ones, which carry sap and fluids and are responsible for the growth pattern.

The **cell walls** are composed of **cellulose** (Figure 11.13), which is a linear crystalline polymer with a degree of polymerization (DP) of ~5,000-10,000. Wood contains about 45-50% cellulose.

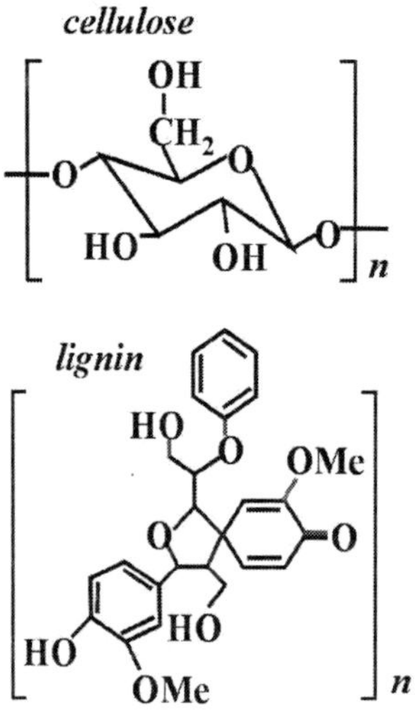

Fig. 11.13 Structures of cellulose and lignin that are essential components of wood.

The cells are held together by a *matrix* of lignin and hemicellulose. Wood contains about 20-25% **hemicellulose**, which is a branched amorphous polymer with a relatively low degree of polymerization (DP ~ 150-200), and about 20-30% **lignin** (a 3D crosslinked phenol-propane network-forming polymer (Figure 11.13)).

Wood reveals a **highly anisotropic** microstructure due to the **tubular cell** alignment. These longitudinal cells serve as very effective **reinforcements** similar to glass fibers in fiberglass.

Note that owing to the highly anisotropic microstructure of wood, the corresponding mechanical properties also show a pronounced anisotropy (chopping wood is easier along the fiber axis than perpendicular to the vertically aligned fibers).

11.2.2 Bone (natural composite)

Bone, nature's premier building material, is a combination of 63 wt% **hydroxylapatite**, $Ca_{10}(PO_4)_6(OH)_2$, with 36 wt% **collagen** (a protein; the amino acid prolin) and about 1 wt% of viscous **liquid**.

Bone is an impressive structural composite, based on its adequate mechanical properties and its ability to **repair** and **remodel** itself. Its microstructure is rather complex, as can be seen in Figure 11.14.

Orthopedic surgeons use the term *large defects*, when, for example, a large portion of bone has to be removed. In former times, such large defects were repaired by harvesting bone from another part of the body (**autografts**, autogeneous bone grafting) or by using cadaver bone (**allografts**). Both of those techniques, however, are accompanied by a number of serious problems for the patient. Therefore, there is a need for **artificial bone replacement** via engineered composites. One example of a man-made product used to fill such large defects is the product Collagraft™.

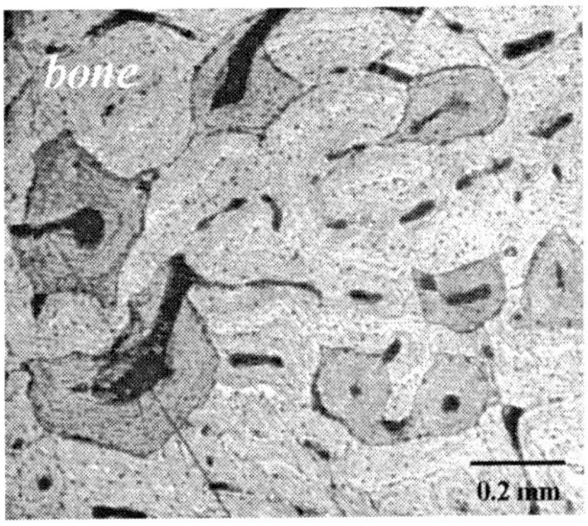

Fig. 11.14 LM image of a cordial bone of a cow (outer dense region); the dark areas represent blood vessels.

In this artificial bone implant system (Collagraft™), millimeter-scale *ceramic particles* are embedded in a collagen matrix. Each ceramic particle is a fine mixture of micrometer-scale grains composed of:

- **hydroxylapatite**, $Ca_{10}(OH)_2(PO_4)_6$, (HAp)
- **tricalcium phosphate**, $Ca_3(PO_4)_2$, (TCP).

Tricalcium phosphate (TCP) is chemically similar to hydroxylapatite (HAp), but reacts more rapidly with the physiological environment. This combination is effective since TCP is **resorbed** by the human body within only a few weeks, allowing rapid attachment of the implant to natural bone. Bone can grow into such pores created by the resorbtion process, while HAp, on the other hand, retains its integrity for a few months, strengthening the patient in the initial period of recovery until all HAp is also resorbed. ***Note*** that both compounds, TCP and HAp, are **biocompatible** (do not cause immune reactions) and **bioactive** (can be resorbed).

11.2.3 Concrete

An excellent example of an **aggregate composite** in which *particles* reinforce the matrix is concrete. In concrete, particle reinforcement is achieved by the addition of **sand** (*fine* aggregate) and **gravel** (*coarse* aggregate) to the cement matrix, which is a mixture of different **Ca-alumosilicates** (see hydration reactions below).

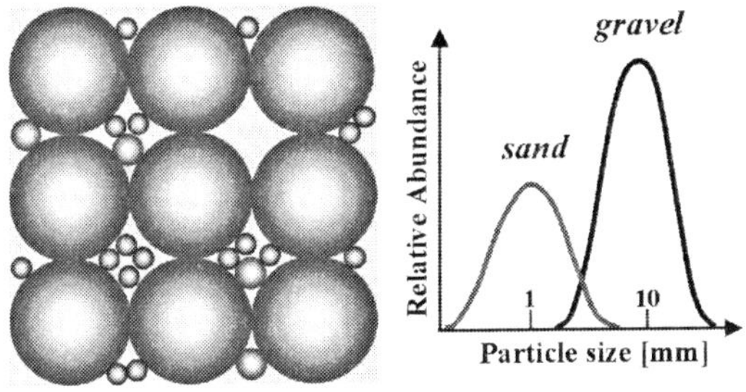

Fig. 11.15 Addition of fine sand and coarse gravel to the cement matrix increases the packing density of the composite.

The reason for choosing a combination of fine and coarse aggregates is to gain a higher **packing density**, based on the bimodal particle size distribution (Figure 11.15). Here, the fine grain fraction can *fill* the *space* between the coarse gravel leading to a packing density of 60-70%, which is much higher than using only one monosized reinforcement fraction. The composition of concrete used in construction applications commonly is:

- 40% coarse aggregate (gravel)
- 25% fine aggregate (sand)
- 20% H_2O
- 10% portland cement
- 5% air.

Portland cement is named after the *Isle of Portland*, Great Britain, where a local limestone closely resembles the composition of the synthetic product (Ca-aluminosilicate). There are five common types of Portland cement, which vary in their composition and their respective properties:

- standard
- increased sulfate resistance *(reduced heat of hydration)*
- high early strength *(high heat of hydration)*
- low heat of hydration *(massive structures)*
- sulfate resistance *(marine structures)*.

The particle size of cement affects the rate at which the cement hydrates. These **hydration reactions** *harden* the cement and produce the chemical bonding between matrix and aggregate particles. In the following, the most important *hydration reactions* are listed:

- $2C_3S + 6H \Longrightarrow 3Ch + C_3S_2H_3$ *($Ca(OH)_2$ plus tobermorite)*
- $2C_2S + 4H \Longrightarrow Ch + C_3S_2H_3$ *($Ca(OH)_2$ plus tobermorite)*
- $C_3A + 10H + CsH_2 \Longrightarrow C_3ACSH_{12}$ *(calcium alumino monosulfate hydrate)*
- $C_3A + 12H + Ch \Longrightarrow C_3ChAH_{12}$ *(tetracalcium aluminate hydrate)*
- $C_4AF + 10H + 2Ch \Longrightarrow C_6AFH_{12}$ *(calcium aluminoferrite hydrate)*

with: $\mathbf{C_3S} = 3CaO{\cdot}SiO_2$, $\mathbf{C_2S} = 2CaO{\cdot}SiO_2$, $\mathbf{C_3A} = 3CaO{\cdot}Al_2O_3$, $\mathbf{C_4AF} = 4CaO{\cdot}Al_2O_3$ Fe_2O_3, $\mathbf{H} = H_2O$, $\mathbf{Ch} = Ca(OH)_2$, $\mathbf{Cs} = CaSO_4$.

Note that the *weight fraction* of concrete annually used in the US exceeds that of all other metals combined. In some cases, concrete is reinforced by steel (rebar, Figure 11.16), which generates *compressive stresses* in the concrete and thereby leads to a higher strength.

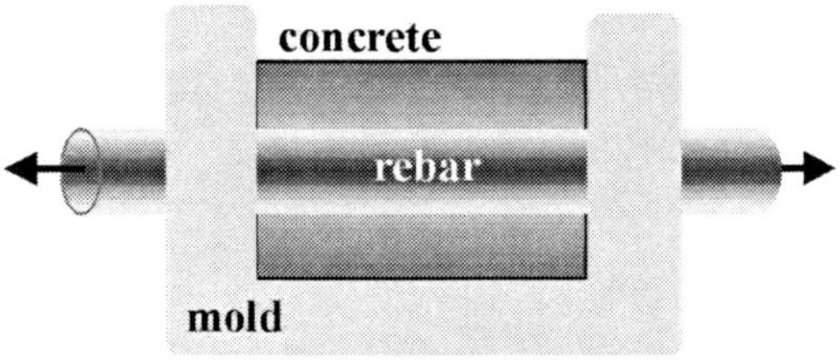

Fig. 11.16 Reinforcement of concrete by rebar which create compressive stresses in the concrete component.

The way the compressive stress is caused in the concrete component is by pre-stressing the rebar, then filling the mold with concrete and letting it harden. Upon hardening, the pre-stress on the rebar is released, resulting in compressive loading of the concrete increasing the strength of the component.

Note that **asphalt** is also an *aggregate composite* composed of bitumen and stones (rocks). Bitumen, a product of petroleum refinement, is a hydrocarbon with additional oxygen and sulfur and behaves mechanically similar to thermoplastic polymers.

11.2.4 High-Tech Composites

Apart from the natural and industrial composites, research over the last three decades has focused on the synthesis and study of composites for specific technical applications. There are quite a number of "High-Tech Composites" with very different characteristics and fields of application; only a few, more general groups are listed below:

- **polymer-matrix** composites (graphite in epoxy or polyester)
- **metal matrix** composites (B/Al, Al/SiC, Mg/Al_2O_3, W/Cu)
- **dispersion-strengthened metals** (ThO_2 precipitates in Ni, Al_2O_3 in Cu)
- **ceramic-metal** composites (Cermets) (Al_2O_3/Fe, SiC/Si, WC/Co SiC/Ti)
- **ceramic-matrix composites** (CMCs) (Al_2O_3/SiC, Si_3N_4/SiC).

An interesting aspect of *dispersion strengthened metals*, which typically contain less than 15 vol% of finely dispersed small particles (precipitates), is the result obtained from rivets from the Titanic, which were recovered in 1986 when parts of the ship hull and some of the rivets were collected from the ocean floor. It was discovered that the rivets contained finely dispersed carbide precipitates that were aligned parallel to the shaft, but revealed a perpendicular orientation in the head region, as shown schematically in Figure 11.17.

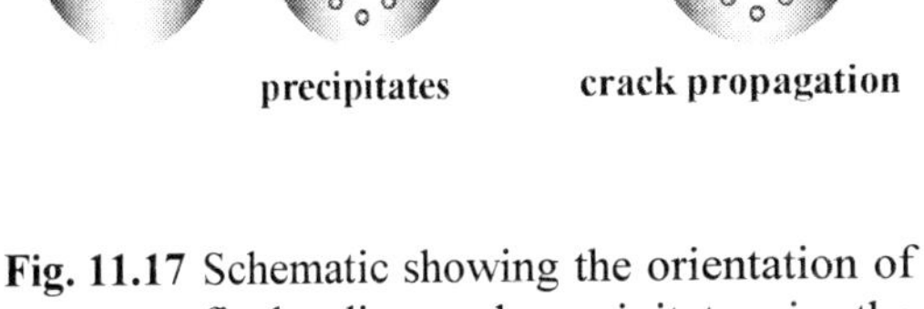

Fig. 11.17 Schematic showing the orientation of finely dispersed precipitates in the shaft and head of rivets recovered from the Titanic wreck.

Due to former processing techniques, the alignment of precipitates within the shaft and the head region of the rivets was different. As a consequence, the heads of the rivets sheared off upon loading. However, this interesting finding did not give a conclusive answer to the overall question why the Titanic sunk and, hence, it is still valid that the main reason for the loss of the Titanic is that it was hit by an iceberg.

11.3 The Interface

In general, the interface is the *key issue* in controlling fracture behavior of composites. It should be emphasized that the interface characteristics play an essential role in determining whether the bonding between the matrix and the reinforcement is **weak** or **strong**.

However, one has to keep in mind that whether an interface is supposed to be weak or strong is mainly dictated by the **application** itself. There are examples where a *strong* interface bond is preferred (strengthening of metals), but more often there are applications where a *weak* interface is wanted. One example of the latter is fiber-reinforced glass ceramics with improved fracture toughness.

Here, we will focus on the discussion of fiber reinforcement of brittle matrices, where the main goal is an increase in fracture toughness. In order to achieve that goal, the following mechanisms have to be operative simultaneously:

- **debonding along** the interface
- **elastic bridging** of the crack wake
- **crack deflection**.

When a propagating crack approaches a fiber which is, for the sake of argument, aligned *perpendicular* to the direction of crack propagation, then the stress field at the crack tip will at some point experience the bond strength of the interface. At that moment, the influence of a strong versus a weak interface bonding on the crack path has to be considered.

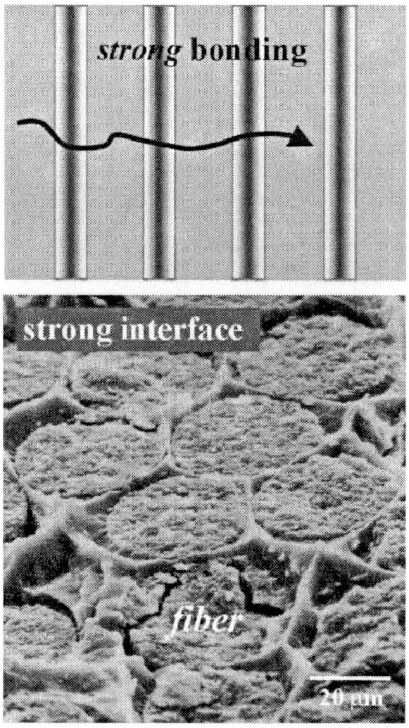

Fig. 11.18 Schematic of crack propagation and SEM image of a fracture surface of a fiber-reinforced composite with a strong bond between matrix and fiber. *Note* that the crack went straight through the fibers, hardly noticing their presence.

Assuming the *bonding* between matrix and reinforcement is **strong** (Figure 11.18), the propagating crack (crack tip) will not be able to differentiate between matrix and fiber. It will only experience a strong bond, similar to the bonding in the matrix, and thus continue to propagate without noticing the fiber. The result is a fairly smooth fracture surface with all fibers fractured. In the situation shown in Figure 11.18, the strong bonding characteristics between fiber and matrix do not affect the crack path and the mechanical behavior of the composite is similar to a monolithic non-reinforced brittle material.

However, when the interface bond strength is low, the propagating crack experiences a **weak** link between reinforcement and matrix (Figure 11.19), which leads to **debonding** of matrix and fiber along the interface.

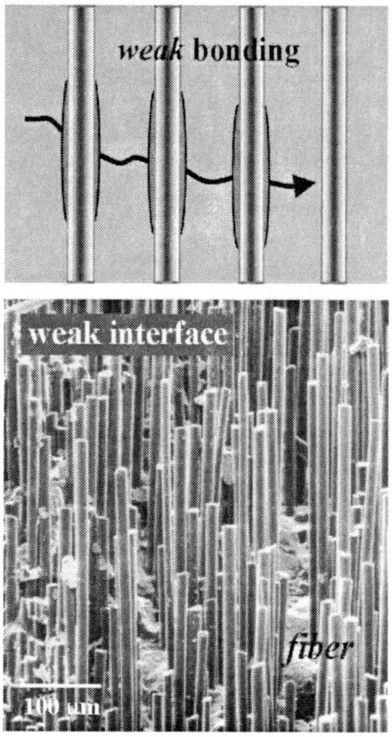

Fig. 11.19 Schematic of the crack propagation path and corresponding SEM image of the fracture surface of a composite with weak interface bonding between fiber and matrix. *Note* that extensive fiber pull-out occurred here.

In the latter scenario, the crack propagates parallel to the interface for a short distance, instead of fracturing the fiber immediately. This *debonding process* continues until the fiber is surrounded by the crack. The crack will then continue to propagate through the matrix, leaving an *intact* fiber behind, as shown in Figure 11.18 (top).

When weak interface bonding is given, debonding along the fiber axis results in **elastic bridges** in the crack wake (unbroken fibers), which act as a force opposing the applied load. Hence, crack propagation becomes more difficult (a higher applied load is needed to drive the crack) and fracture toughness increases considerably. In addition to the fibers bridging the crack wake, **fiber pull-out** occurs; a process that dissipates energy via friction (generation of heat).

Another mechanism that can contribute to the increase in fracture toughness, although to a lesser degree, is **crack deflection**, shown in Figure 11.20.

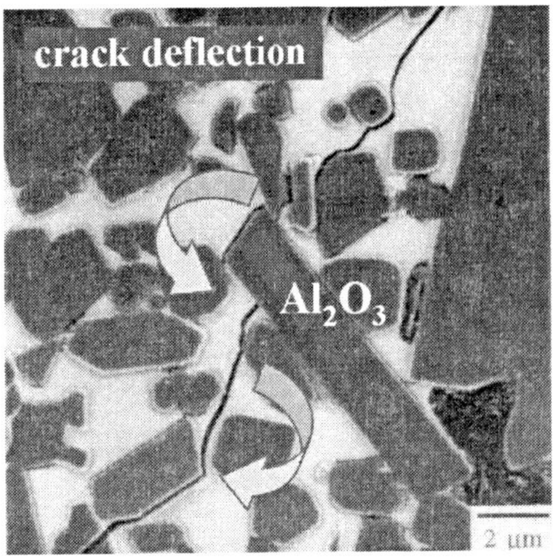

Fig. 11.20 SEM image of a crack path in a ceramic matrix composite revealing debonding at particle matrix interfaces and crack deflection.

In particular, when the direction of crack propagation is *not perpendicular* to the fiber, the crack path follows the direction of the weak interface, causing an increase in overall crack length.

Note that for some fiber-reinforced composites with weak bonding between matrix and fiber, an improvement in fracture toughness from 3 to 20 $MPa(m)^{1/2}$ was reported (Figure 11.21).

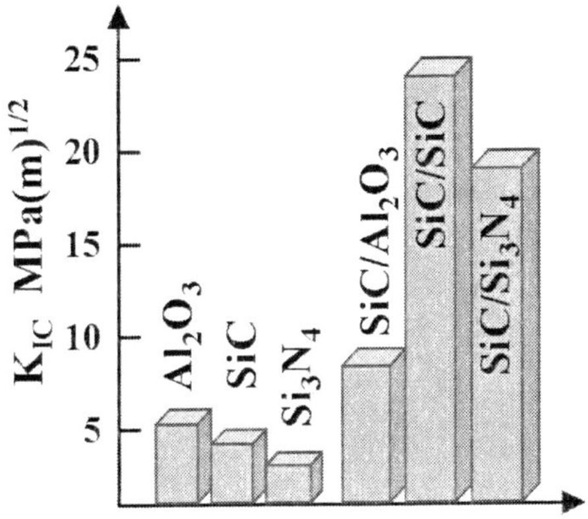

Fig. 11.21 Comparison between the fracture toughness of non-reinforced and fiber-reinforced structural ceramics.

Questions

What is the principle of the rule of mixtures?

How is the term 'anisotropy' defined?

What are the constituents of natural bone?

Fiber-reinforced brittle matrices require weak interface bonding to improve fracture toughness; why?

12 Thermal Properties

In this chapter we will discuss the thermal properties of various materials, ranging from ceramics and glasses to metals.

Heat capacity and **specific heat** indicate a material's ability to *absorb heat* from the environment. The energy imparted to the material from an external heat source produces an increase in *thermal vibration* of the atoms in the material (similar to sitting in a bus that runs along a bumpy road).

Most materials show a slight *increase in volume*, when externally heated. This **thermal expansion** is a direct consequence of the greater separation distance between the centers of adjacent atoms with increasing temperature.

When the flow of heat through a given material is described, the **thermal conductivity** is the proportionality constant between the heat flow rate and the temperature gradient. In similar terms, we defined *diffusivity* as the proportionality constant between the *mass flow rate* and the *concentration gradient* (Fick's laws).

Refractories are high-temperature resistant materials such as ceramics (mullite) that are used, for example, in metal casting applications. The most *effective* refractories have *low thermal expansion* and *low thermal conductivity* values.

When materials experience rapid changes in temperature (either via heating or cooling), thermal stresses can be induced at the material surface, depending on the flow of heat within the material. **Thermal shock** refers to the fracture of a material that was exposed to a rapid temperature change, similar to dropping a hot glass container into a bucket of ice water, which is termed *quenching*. It follows a more detailed discussion of those properties that define the *thermal response* of a material being exposed to high service temperatures.

12.1 Heat Capacity

As a material is subjected to heat, its temperature rises, since heat is absorbed by the material. This phenomenon can be quantified in terms of the material property **heat capacity**, C, defined as the amount of heat required to raise the temperature of the material by 1 K (1 °C):

$$C = \frac{Q}{\Delta T} \tag{12.1}$$

with Q being the amount of heat that produces the temperature change, ΔT. The magnitude of C will depend on the amount/volume of the material. The heat capacity is commonly specified for either one gram-atom or one mole:

- one gram-atom (element) [J/g-atom·K]
- one mole (compound) [J/mol·K].

12.2 Specific Heat

The **specific heat** (written in lowercase letters) is defined as the amount of heat per unit mass, m, to raise the temperature by one degree [J/kg·K]:

$$c = \frac{q}{m\Delta T} \tag{12.2}$$

Note that the specific heat of water at constant atmospheric pressure is 1.000 cal/g·K = 4.184 J/g·K.

There are two ways to experimentally determine the heat capacity or the specific heat:

- under constant volume $C_v (c_v)$
- under constant pressure $C_p (c_p)$.

Note that the magnitude of $C_p (c_p)$ is always greater than $C_v (c_v)$; however, the difference is only minor for most solids at room temperature or below. Commonly, data for atmospheric pressure is used, as given in Table 12.1.

Tab. 12.1 Specific heat values of some common metals, ceramics, and polymers.

Specific Heat [J/kg·K]				
Material	**Au**	129	**Al_2O_3**	160
	W	133	**SiC**	344
	Pb	159	**MgO**	457
	Ag	237	**$C_{diamond}$**	519
	Cu	385	**$C_{graphite}$**	711
	Fe	444		
	Ni	444	**PTFE**	1050
	Ti	523	**Nylon66**	1400
	Al	900	**PP**	1800
			PE	2000

Fundamental studies in the early 1920's and 1930's, which established the correlation between atomic vibration and heat capacity, showed that C_v rises sharply above zero Kelvin, following equation 12.3:

$$C_v = AT^3 \qquad (12.3)$$

where A is a temperature-dependent constant. As shown in Figure 12.1, the heat capacity first rises sharply and then, above the Debye temperature [30], T_D, C_v tends to level off at a value of approximately $3R$, with R being the universal gas constant (the Debye temperature, T_D, is the highest temperature of a crystal due to a single normal vibration).

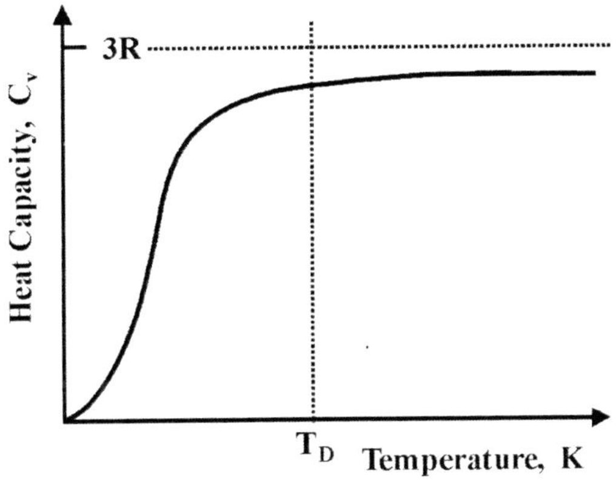

Fig. 12.1 Temperature dependence of the heat capacity (constant volume). *Note* that above the Debye temperature, T_D, the heat capacity, C, asymptotically approaches the value of 3R.

12.3 Thermal Expansion

An increase in temperature leads to a greater thermal vibration of the atoms within the material. As a consequence, a greater separation distance of adjacent atoms is generated with rising temperature, resulting in an increase in sample dimension (dL).

The *linear coefficient* of **thermal expansion**, α, is given by:

$$\alpha = \frac{dL}{L \cdot dT} \qquad (12.4)$$

with α having the dimension [mm/mm·°C].

[30] **Peter Joseph Wilhelm Debye** (1884-1966), Dutch-American physical chemist.

Thermal expansion is generally smaller for ceramics than for metals, which is again smaller for polymers. The difference in the thermal expansion coefficient of different materials is related to the **asymmetry** of the *bonding energy well* (bond energy versus separation distance), as shown in Figure 12.2.

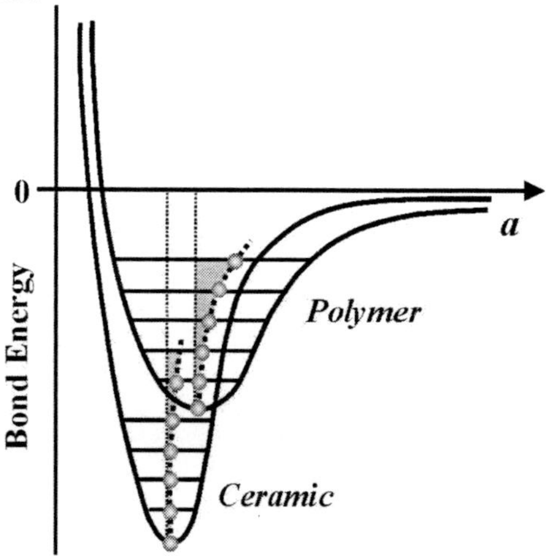

Fig. 12.2 Bond energy versus separation distance, a, for polymers versus ceramics. *Note* that ceramics show a deeper and less asymmetric energy well (lower thermal expansion, α).

Ceramics and glasses have a rather *deep* **energy well** due to their mixed bonding type (ionic and/or covalent)) and, therefore, reveal a more symmetric bonding energy curve than for example polymers. ***Note*** that the *larger/deeper* the energy well, the *stronger* the bonding and the *higher* the *elastic modulus* (Young's modulus). Moreover, it follows that a deep energy well results in high melting temperatures, T_m, but a lower thermal expansion coefficient, α:

- **weak bonding**
  - ❑ low melting temperature
  - ❑ low elastic modulus
  - ❑ high thermal expansion
- **strong bonding**
  - ❑ high melting temperature
  - ❑ high elastic modulus
  - ❑ low thermal expansion.

Thermal expansion, α, is a function of temperature, as shown in Figure 12.3. Commonly, the coefficient of thermal expansion increases with temperature, but levels off at higher temperatures.

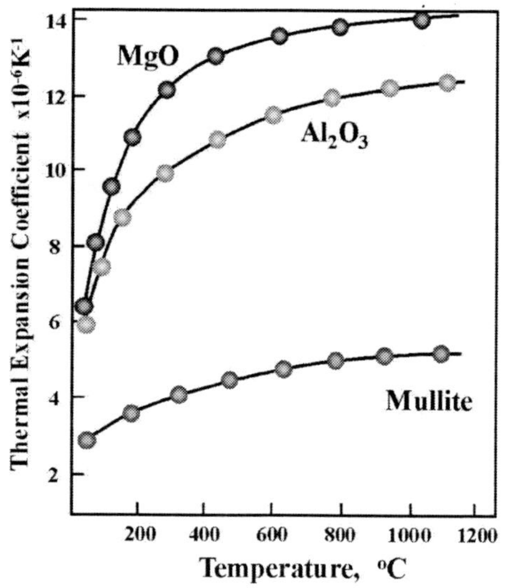

Fig. 12.3 Thermal expansion coefficient versus temperature for three common ceramic materials.

Note that some solids reveal a *negative* coefficient of thermal expansion such as **β-eucryptite**, $Li_2O \cdot Al_2O_3 \cdot 2SiO_2 = LiAlSiO_4$. With rising temperature, the crystal structure of β-eucryptite relaxes, resulting in a slightly higher density (reduction in volume). Such materials are used to produce components with an overall *low thermal expansion* such as glass ceramics (cookware), which therefore have an *excellent thermal shock resistance*.

12.4 Thermal Conductivity

Heat conduction within solids can be described as being analogous to the *mass transport* in solid compounds. The analog for the diffusivity, D, during mass transport is the **thermal conductivity**, k, during heat conduction. Thermal conductivity is defined by **Fourier's law** [31] as:

$$k = -\frac{dQ/dt}{A \cdot (dT/dx)} \qquad (12.5)$$

where dQ/dt equals the heat transfer rate across the area, A, due to the presence of a temperature gradient, dT/dx (Figure 12.4). The latter acts as the **driving force** similar to the compositional gradient, dc/dx, that acts as the driving force for cation/anion diffusion (Fick's law).

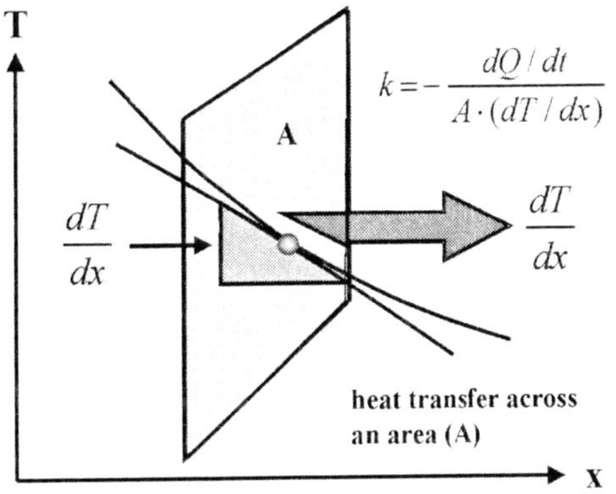

Fig. 12.4 Schematic illustrating Fourier's law of heat transfer across an area, *A*.

For a steady-state heat conduction through a flat slab, the differentials in equation 12.5 become average terms:

$$k = -\frac{\Delta Q/\Delta t}{A \cdot (\Delta T/\Delta x)} \qquad (12.6)$$

[31] **Jean Baptiste Joseph Fourier** (1768-1830), French mathematician.

Similar to the thermal expansion coefficient, α, thermal conductivity, k, is also a *function of temperature*. For some materials, thermal conductivity increases with increasing temperature, while other materials show the opposite response (Figure 12.5). Which trend the thermal conductivity of a given material follows depends on the *mechanism of heat conduction* (see below) and on other variables such as the volume fraction of residual porosity.

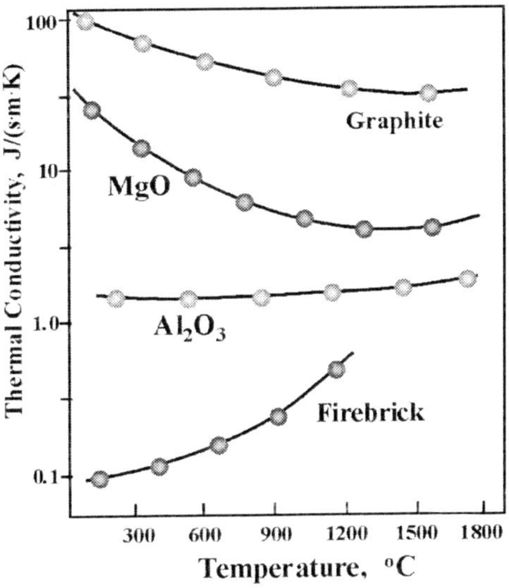

Fig. 12.5 Temperature dependence of thermal conductivity of ceramic materials.

12.4.1 Mechanisms of Thermal Conduction

The conduction of heat in materials involves *two primary mechanisms*:

- **atomic vibration**
- **conduction of free electrons.**

For *insulators* (poor electrical conduction) such as ceramics and glasses as well as most polymers, thermal energy is primarily conducted via **atomic vibration.**

For good *electrical conductors* such as metals and alloys, the *kinetic energy* of the **conducting** (free) **electrons** provides a much more efficient heat conduction mechanism than atomic vibration, which however occurs simultaneously.

Note that **electron dualism** implies that an electron can be viewed as a *particle* (SEM imaging) as well as a *wave* (HRTEM imaging). For any given wave, a *structural disorder* within the material will *interfere* with the propagating wave. Hence, with rising temperature and increasing thermal (atomic) vibration of the atoms on their lattice sites within the crystal structure, thermal conductivity *decreases*.

As a consequence, any **structural disorder** within the crystal will cause a *decrease* in thermal conductivity. Therefore, *alloys* tend to have a lower thermal conductivity than the corresponding pure metals.

For ceramics and polymers, *thermal vibrations* are the predominant mechanism of thermal conduction. Their collective lattice vibrations are also wavelike in nature and are, therefore, similarly affected by structural disorder. As a result, *glasses* tend to have a lower thermal conductivity than the corresponding crystalline ceramics; the same is true for polymers:

$$k_{amorph} \angle k_{crystalline} \qquad (12.7)$$

Note that thermal conductivity of polymers and ceramics is lowered with increasing temperature, due to the increase in structural disorder, a consequence of the increased atomic vibration. One can view it as an increase of the "virtual" size of the atoms with rising temperature, acting as larger obstacles.

Porosity will *further reduce* the thermal conductivity of all materials, since the gas that is entrapped in the pores typically has a rather low value of thermal conduction. Two well-known examples of materials with low thermal conductivity, both containing residual porosity, are:

- **polystyrene drinking cups**
- **space shuttle tiles.**

12.5 Thermal Shock

The use of inherently brittle materials such as ceramics and glasses at *high and low* temperatures causes a specific engineering problem, commonly referred to as **thermal shock resistance.** One can visualize this behavior as the partial or catastrophic failure of the material upon a sudden temperature drop (quench).

The **mechanism** affecting thermal shock behavior involves both *thermal expansion*, α, and *thermal conductivity*, k. The salient feature is a rapid decrease in retained strength around a critical temperature difference, ΔT_c, below which the original strength is retained (see Figure 12.6).

For practical purposes (application of a given component at various temperatures) it is important to be able to predict the critical value of ΔT_c, above which a pronounced decrease in retained strength is observed (Figure 12.6). Moreover, only when the various parameters that affect thermal shock behavior are understood is it possible to design a material that can withstand a drastic and rapid temperature change.

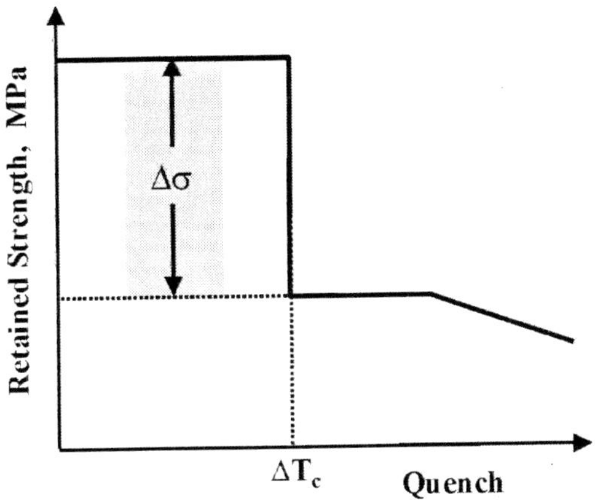

Fig. 12.6 Schematic of strength behavior as a function of the magnitude of the quench, ΔT.

When the quench temperature is further increased, the strength further decreases, but more gradually.

To estimate ΔT_c, a number of assumptions are commonly made:

- the material contains N identical, uniformly distributed Griffith flaws per unit volume
- the body is uniformly cooled with the external surfaces rigidly constrained to give a well-defined tensile stress state
- crack propagation occurs by the simultaneous propagation of these N cracks, with negligible interaction between the stress fields of neighboring cracks.

The maximum stress at the surface, σ_{surf}, that is generated in a relatively thick body when quenched into a medium (ice water) is given by:

$$\sigma_{surf} = \frac{E \cdot \alpha}{1-\nu} \cdot (T_a - T_s) \qquad (12.8)$$

where E is the elastic modulus, α is the thermal expansion coefficient, ν the Poisson ration of the material, T_a is the average temperature of the body, and T_s the surface temperature. In the worst-case scenario, the average temperature equals the highest temperature inside the sample and the surface temperature equals the temperature of the cooling medium. Then, $(T_a - T_s) = \Delta T_c$ and by rearranging equation 12.8 it follows:

$$\Delta T_c = \frac{\sigma_f \cdot (1-\nu)}{E \cdot \alpha} \qquad (12.9)$$

where σ_f is the failure strength of the material. It follows that a material that can withstand a rapid quench from a high temperature should show the following characteristics:

- low elastic modulus
- low thermal expansion
- low Poisson ratio
- high fracture strength.

A rapid temperature change will produce a **temperature gradient** within the material, which in turn generates *tensile stresses* at the sample surface (Figure 12.7), because the surface contracts more rapidly than the interior.

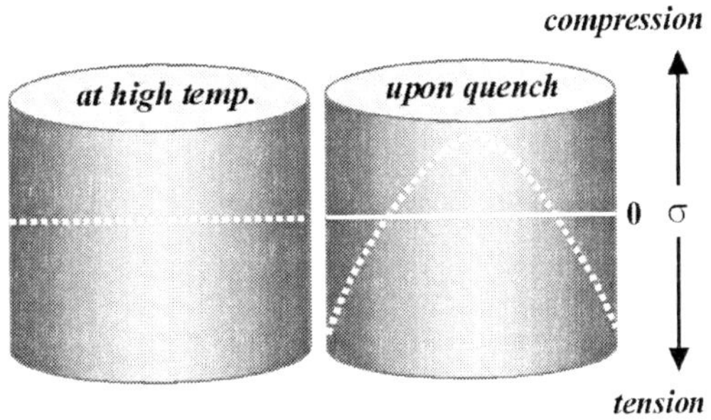

Fig. 12.7 Surface stress state of a ceramic body upon quenching.

These tensile surface stresses have the potential to lead to brittle fracture, since they typically generate small surface flaws (Griffith flaws). For example, in un-stabilized *monoclinic* ZrO_2 even small temperature changes will cause cracking. *Soda-lime glass* will not withstand a rapid temperature change (quench) of 100 °C, while *fused silica*, due to it's low thermal expansion coefficient of $0.5 \times 10^{-6}/^{o}C$, can survive a quench from 1600 °C (space shuttle tiles).

In order to avoid such critical surface stresses (in particular when designing with ceramics and glasses), it is important to carefully match thermal expansion coefficients and the elastic moduli of the different materials that are in direct contact (for example blending of artificial teeth with an enamel coating).

Note that even without external constraints, thermal shock can occur due to temperature gradients created by the finite nature of thermal conductivity.

Questions

What is the definition of heat capacity?

Why is the thermal expansion coefficient of a ceramic lower than of a polymer?

What would be the analogous to Fourier's law (heat transfer) for mass transport?

What are the main mechanisms of thermal conduction?

Why does fused silica show an excellent thermal shock resistance?

13 Optical Properties

In general, optical properties of materials can be described as a material characteristic relative to the **interaction with light**. Optical properties of materials play an important role in a number of today's high technologies (optical fibers, lasers).

Visible light is one form of **electromagnetic radiation** with a wavelength, λ, extending from about 400 to 750 nm, containing different color bands (Figure 13.1). The *ultraviolet* region covers the range of about 0.01 to 0.40 µm, while the *infrared* region of the electromagnetic spectrum of light extends form about 0.75 to 1000 µm.

Light can be considered to have a *wave* form and to consist of *particles* termed **photons**. The energy, E, wavelength, λ, and frequency, ν, of photons are related by the fundamental equation:

$$E = h \cdot \nu = \frac{h \cdot c}{\lambda} \qquad (13.1)$$

where h is Planck's constant, 6.62×10^{-34} Js, and c is the speed of light in vacuum, 3.00×10^{8} m/s. Equation 13.1 describes a photon as a particle of energy E and/or as an electromagnetic wave with a given wavelength and frequency.

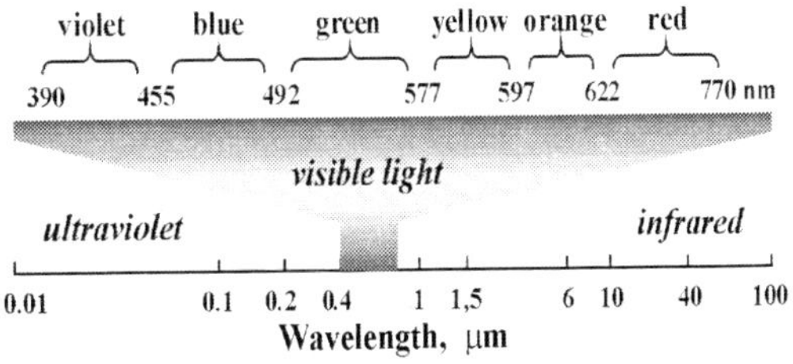

Fig. 13.1 Electromagnetic spectrum of light extending from the ultraviolet to the infrared.

In the following sections, we will discuss the ability of materials to *transmit* or *reflect* light.

13.1 Refractive Index

One of the most fundamental optical properties of materials is the **refractive index**, n, which is defined as:

$$n = \frac{v_{vac}}{v_{mat}} = \frac{\sin \Theta_i}{\sin \Theta_r} \qquad (13.2)$$

where v_{vac} is the speed of light in vacuum and v_{mat} is the velocity of light in the medium considered, while Θ_i and Θ_r are the angles of incidence and reflection, respectively, as given in Figure 13.2.

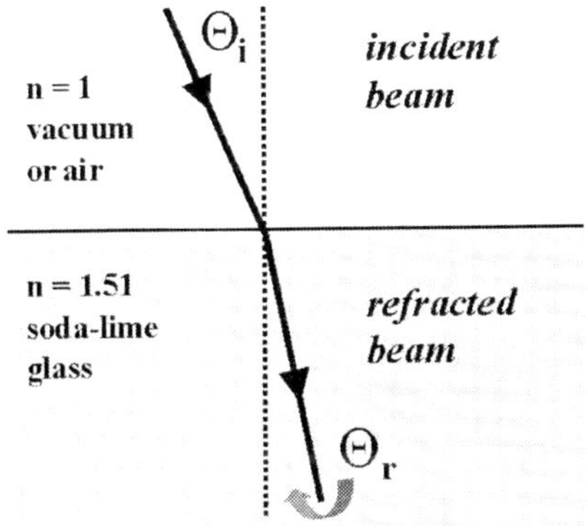

Fig. 13.2 Schematic showing the refraction of a light beam passing from vacuum (or air) into a medium with higher refractive index, n.

When light photons are transmitted through a transparent material, they lose some of their initial energy. As a result, the speed of light is reduced and the beam of light **changes direction**.

Note that the beam of light is refracted towards the material with higher refractive index; $\Theta_i > \Theta_r$ (compare Figure 13.2). Moreover, the refractive index of a material is different for different wavelengths of the incoming light.

Typical *average* indices of refraction for some glasses, polymers and crystalline solids are shown in Table 13.1. Their values range from 1.4 to 2.6, with a refractive index of 1.5 to 1.7 for most silicates.

Tab. 13.1 Refractive indices of glasses, crystalline solids, and some polymer materials.

Refractive Index, n	
SiO_2 glass	**1.46**
Borosilicate glass	**1.47**
Soda-lime glass	**1.51-1.52**
SiO_2, quartz	**1.55**
$Al_6Si_2O_{13}$, mullite	**1.64**
$MgAl_2O_4$, spinel	**1.72**
MgO, periclase	**1.74**
Al_2O_3 corundum	**1.76**
C, diamond	**2.41**
PbO, litharge	**2.61**
Polystyrene, PS	**1.59-1.60**
Polyethylene, PE	**1.50-1.54**
Polymethyl meta-acrylate, PMMA	**1.48-1.50**
Polytetrafluoro-ethylene, Teflon	**1.30-1.40**

If the beam of light passes from one medium with a *high* refractive index n into another medium with a *low* refractive index n', their refractive indices are related with the incident angle, Θ_i, and the refractive angle, Θ_r, by **Snell's law** [32] of refraction:

$$\frac{n}{n'} = \frac{\sin\Theta_r}{\sin\Theta_i} \tag{13.3}$$

Note that when light passes from a medium with a *high* refractive index to one with a *low* index of refraction, there is a critical angle of incidence, Θ_c, at which (and at higher values) **total internal reflection** of light will occur at the interface, as shown in Figure 13.3.

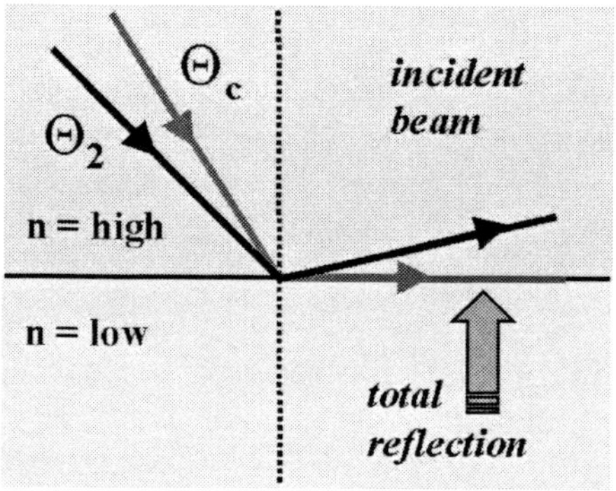

Fig. 13.3 Ray diagram indicating total internal reflection at the critical angle of incidence, Θ_c.

13.2 Absorption, Transmission and Reflection

All materials **absorb light** at least to some degree, due to the *interaction* of the light photons with the *bonding* and *electronic nature* of the atoms, ions, or molecules present in the structure (amorphous or crystalline).

Thus, the fraction of light transmitted by a particular material depends on the amount of light reflected and absorbed. For a special wavelength, λ, the sum of fractions

[32] **Willebrord van Roijen Snell** (1580-1626), Dutch mathematician.

of incoming light transmitted, f_t, absorbed, f_a, or reflected, f_r, equals 1:

$$f_r + f_a + f_t = 1 \qquad (13.4)$$

Not all light striking a transparent **glass ceramic**, for example, enters the material and is *refracted*, as discussed above. A fraction of the light is **reflected** at the *surface* of the material. The **reflectivity** (or reflectance), R, is defined as the fraction of light that is reflected at such an interface and is related to the refraction index, n, by **Fresnel's law** [33]:

$$R = \left(\frac{n-1}{n+1}\right)^2 \qquad (13.5)$$

Note that equation 13.5 is strictly valid only for *normal incidence*, Θ_i = 90°, but represents a good approximation over a wide range of incident angles.

Example 13.1

The reflectivity of a polished silica glass surface with a refractive index of n = 1.46 (see also Table 13.1) can be calculated using Fresnel's law:

$$R = \left(\frac{n-1}{n+1}\right)^2 = \left(\frac{1.46-1.00}{1.46+1.00}\right)^2 = 0.035$$

hence, the reflectivity of a **SiO_2-glass** surface is only:

$$R(100\%) = 0.035 \text{ x } 100\% = \mathbf{3.5\%}.$$

Note that diamond with a higher refractive index of n=2.41 shows a much higher reflectivity of R=17%.

[33] **Augustin Jean Fresnel** (1788-1827), French physicist.

Glass **absorbs** energy from the transmitting light so that the light intensity decreases as the light path increases. The relation between the intensity of incident light, I_0, and the fraction of exiting light at the bottom of a glass plate is given by:

$$\frac{I}{I_0} = e^{-\alpha \cdot t} \qquad (13.6)$$

with t being the glass plate thickness and α the linear absorption coefficient.

The amount of incident light transmitted through a glass plate is determined by the amount of light **reflected** from both *upper* and *lower surfaces* as well as the amount of light that is **absorbed** within the glass plate (Figure 13.4). The fraction of incident light reaching the lower surface of the glass is:

$$(1-R)(I_o \cdot e^{-\alpha \cdot t}) \qquad (13.7)$$

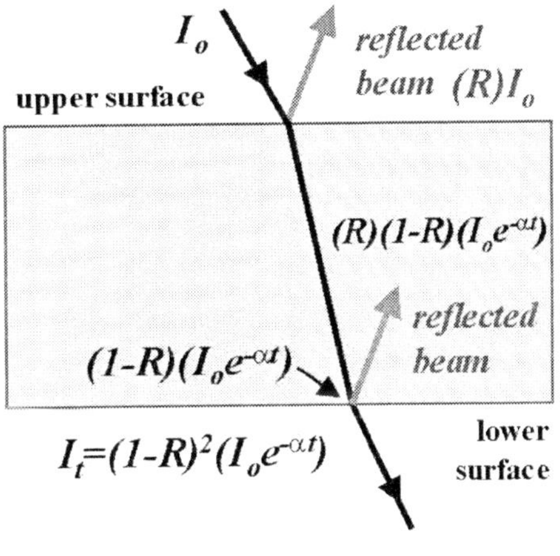

Fig. 13.4 Light transmittance through a glass plate where reflectance occurs at the upper and lower glass surface and absorption takes place within the glass.

The fraction of incident light *reflected* from the *lower surface* is therefore:

$$(R)(1-R)(I_o \cdot e^{-\alpha \cdot t}) \qquad (13.8)$$

and the difference between the light reaching the lower surface of the glass plate and that which is reflected at the lower glass surface is the fraction of light transmitted, I_t:

$$I_t = [(1-R)(I_0 \cdot e^{-\alpha \cdot t})] - [(R)(1-R)(I_0 \cdot e^{-\alpha \cdot t})]$$
$$= (1-R)^2 (I_0 \cdot e^{-\alpha \cdot t}) \qquad (13.9)$$

As shown in Figure 13.5, about 90% of the incoming light is transmitted through a thin silica glass plate. ***Note*** that this is only valid for wavelengths greater than 300 nm. For shorter wavelengths (ultraviolet light) absorption is much higher and transmittance is lowered considerably.

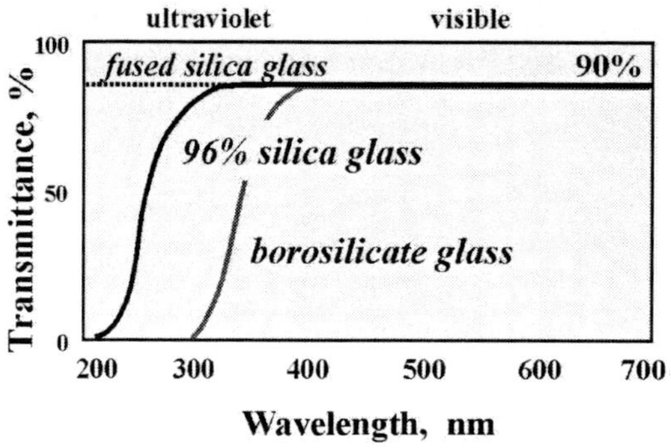

Fig. 13.5 Percentage of light transmittance versus wavelength (nm) for silica and borosilicate glass.

Metals on the other hand, with the exception of very thin sections, show a rather **high reflectivity** and/or **absorb** the incident radiation from long wavelengths (radio waves) to the middle of the ultraviolet range.
Since, in metals, the conduction band overlaps with the valence band, incident radiation can easily **elevate electrons** from lower energy levels to higher levels. When the electrons drop to lower energy levels, the emitted photon energies are low and their wavelengths are long. This type of interaction results in a *strong reflectance* of the incoming light from smooth metal surfaces. The amount of light absorbed by the metal depends on the corresponding electronic structure.

For copper and gold, there is a greater absorption of shorter wavelengths of blue and green and a greater reflection of red, orange, and yellow. Hence, smooth surfaces of those metals show the *reflected colors*. Other metals such as aluminum or silver strongly reflect all parts of the visible spectrum of light and, therefore, show a *white silvery* color.

Note if light is shone on a metal surface, electrons can be released. These electrons can be attracted towards a positively charged metal plate a certain distance below the metal surface, thereby establishing a *photoelectric current*. This **photoelectric effect** was discovered by **Albert Einstein** [34] in 1905. In 1921 he was awarded the Nobel Prize in physics for the discovery of the photoelectric effect (not for his theory on relativity). The same effect can be observed in semiconductors (also see below), when the photon energy is on the order of the band gap, E_g:

$$\lambda = h \cdot c / E_g \qquad (13.10)$$

Many *amorphous* **polymers** such as polymethyl methacrylate (PMMA) or polystyrene (PS) have **high transparency**. However, when those polymers contain small crystalline regions of higher refractive index than the non-crystalline matrix, light waves can be scattered by reflection and refraction (Figure 13.6).

[34] **Albert Einstein** (1879-1955), German physicist.

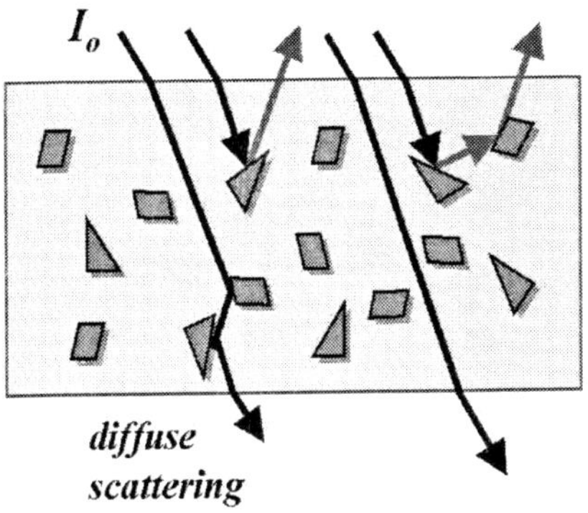

Fig. 13.6 Reduced transparency of polymers due to reflection and refraction of light at small crystalline particles embedded in the amorphous matrix.

When the particle size is larger than the wavelength of the incoming light, the transparency of the polymer is strongly reduced. For example, high-density polyethylene (HDPE) reveals a higher crystallinity than low-density polyethylene (LDPE), which is predominantly amorphous, consequently HDPE is less transparent.

13.3 Semiconducting Materials

In **semiconductors**, light waves can be absorbed in several ways. In **intrinsic** (pure) semiconductors such as Si or Ge, photons are absorbed and create **electron-hole pairs** by causing the electron to jump across the band gap, E_g, from the valence band into the conduction band (Figure 13.7). In order for this mechanism to occur, the incoming photon energy must be equal or greater than the energy gap, E_g. If the energy of the light photon is greater than E_g, the excess energy is *dissipated* as *heat*. For **extrinsic** semiconductors containing **acceptor** or **donor** impurities, lower photon energies are required to excite electrons. Photons are absorbed to either lift an electron from the valence band to an acceptor level (p-type) or from a donor level to the conduction band (n-type; Figure 13.8).

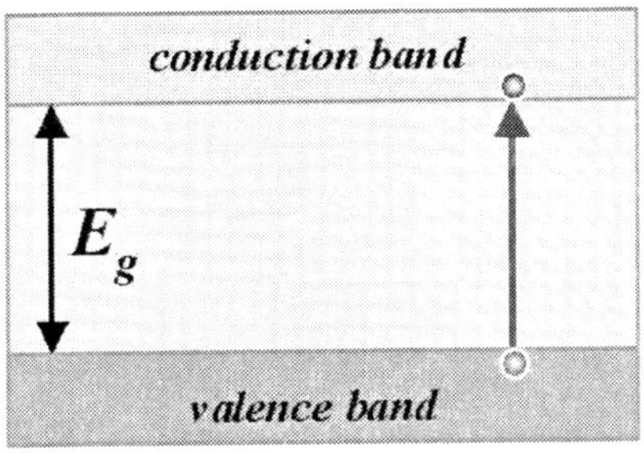

Fig. 13.7 Absorption of photons by an intrinsic (pure) semiconductor.

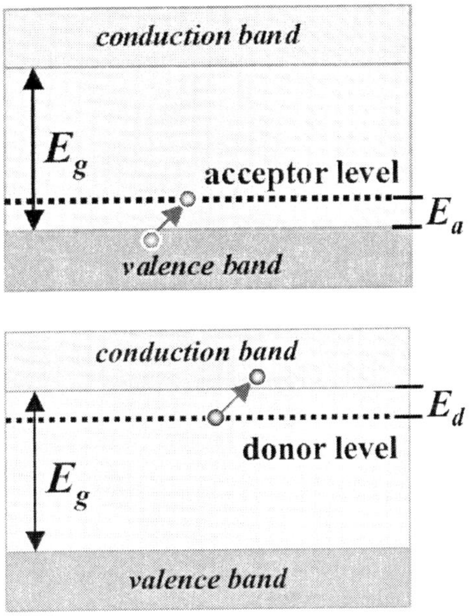

Fig. 13.8 Absorption of light by an extrinsic (doped) semiconductor; n-type (top) and p-type (bottom).

Semiconductors are **opaque** to high and intermediate energy photons (short and intermediate wavelengths) and are transparent to low energy photons with high wavelengths (infrared).

13.4 Luminescence

Luminescence can be defined as the process by which a material **absorbs energy** and then spontaneously **emits visible** or near-visible **radiation**. During this process, the input energy excites electrons of a *luminescent* material from the valence band into the conduction band. The source of input energy may be, for example, high-energy electrons (television) or light photons (pulsed ruby laser).

During **luminescence**, the excited electrons drop to lower energy levels and, in some cases, recombine with holes. Depending how long it takes for the emission to take place, one distinguishes between:

- **fluorescence** where photon emission occurs *within* $\mathbf{10^{-8}}$ seconds
- **phosphorescence** where it takes *more than* $\mathbf{10^{-8}}$ seconds for photon emission to occur.

Note that *fluorescence* is often used in clothing in order to brighten the fabric (emitted blue color masks the yellow color of the fabric), while *phosphorescence* is important for picture averaging in television screens.

Luminescence is produced by materials named **phosphors** which have the capability to absorb energy and *spontaneously emit* light radiation. The emission spectra are commonly controlled by adding activator impurities to the material, which provide discrete energy levels between the valence and the conduction band.

Luminescence processes are also classified according to the energy source used for electron excitation. The two industrially important types are:

- **photoluminescence**
- **cathodoluminescence**
- **electroluminescence**.

During **photoluminescence**, one kind of light is converted to another fluorescent light. In the fluorescence lamp, ultraviolet radiation from a low-pressure mercury arc is converted into visible light using a phosphor coating with the composition $Ca_{10}F_2P_6O_{24}$ (about 20% of the F^-- ions are replaced by Cl^--ions). Additions of Sb^{3+} or Mn^{2+} can change the color of the emitted light between blue and white. The high-energy ultraviolet light from the excited mercury atoms causes the phosphor-coated inner wall of the fluorescent lamp to emit visible light (of lower energy and longer wavelength), as shown in Figure 13.9.

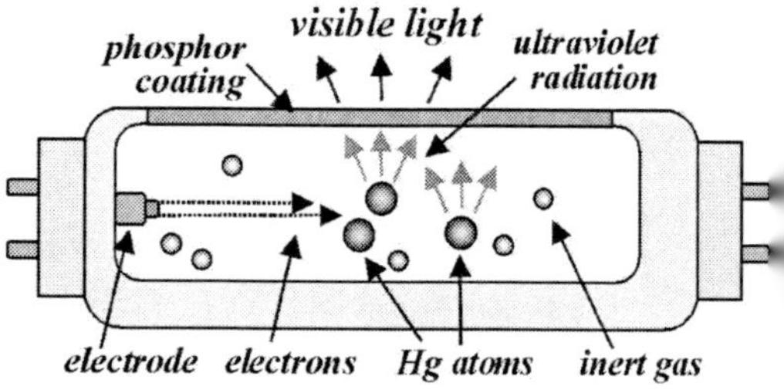

Fig. 13.9 Schematic of a fluorescent lamp where the ultraviolet light emitted from excited mercury atoms causes the phosphor coating to emit visible light.

The principle of **cathodoluminescence** is the conversion of photons into an electrical signal (*or the reverse*) and is used in a number of technical applications such as the scanning electron microscope (SEM), cathode-ray oscilloscope, and color TV-screen; the latter is shown in Figure 13.10.

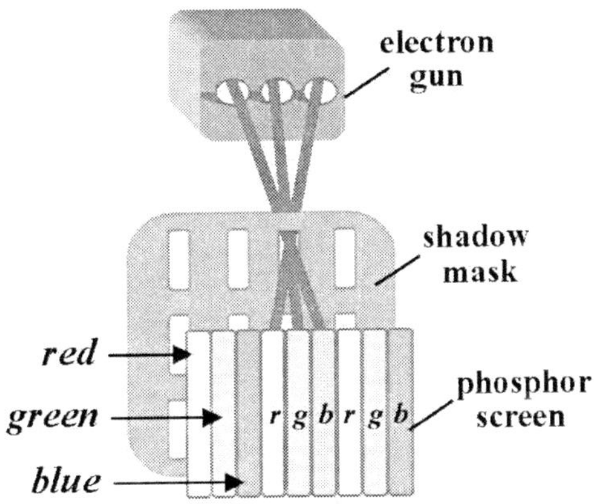

Fig. 13.10 Schematic illustrating the principle of cathodoluminescence used in a color TV-screen.

Here, a beam of high-energy electrons is used to produce luminescence. The process of color TV **phosphorescence** involves narrow dots or stripes of red, green, and blue emitting phosphors and a steel shadow mask with small, elongated holes. The high-energy electron beam is rapidly scanned over the entire screen and the desired color from each "three-color set" is emitted. Commonly used phosphors for the three colors are:

- **blue**, ZnS with Ag^+-acceptor and Cl^--donor addition
- **green**, (Zn,Cd)S with Cu^+-acceptor and Al^{3+}-donor additions
- **red**, Y_2O_2S with ~3% europium (Eu) addition.

The phosphor materials must retain some image glow until the next electron-beam scan, but should not glow too long to blur the image. The intensity of luminescence, I_l, is given by:

$$\ln\frac{I_l}{I_0} = -\frac{t}{\tau} \qquad (13.11)$$

where I_0 is the initial intensity of luminescence, I_l is the fraction of luminescence after time t, and τ is the *relaxation time constant* of the phosphor.

Electroluminescence occurs in materials that are both *phosphors* and *semi-conductors*. A well-known example is **light-emitting diodes** (LED's). This process refers to the production of light when an **electric field** is applied to a solid body and was first discovered in **silicon carbide**. LED's are commonly employed as display elements in electronic calculators and digital watches.

13.5 Lasers

Laser is an acronym that stands for "*light amplification by stimulated emission of radiation*". ***Note*** that conventional light sources emit photons of similar wavelength *independently* and *randomly*. As a consequence, this radiation is emitted in random directions and the different waves are out-of-phase with each other, which is termed **incoherent** radiation.

In contrast, a *laser* light source produces a photon emission that is **coherent**, meaning it is in-phase, highly directional, and (nearly) monochromatic. A ruby laser (Al_2O_3 doped with ~0.05% Cr^{3+}) uses a xenon flash lamp as high-energy input signal to excite the Cr^{3+}-ion electrons (fluorescent centers) in the Al_2O_3 lattice to a higher energy level. This process is termed **pumping** the laser. The excited electrons of the Cr^{3+} may then drop back to a lower energy level, either to the ground state or to a *metastable* energy state. However, before the stimulated emission of photons can occur in a laser material, more electrons have to be *pumped* into the metastable energy level than are present in the ground state. This process is referred to as **population inversion** (Figure 13.11).

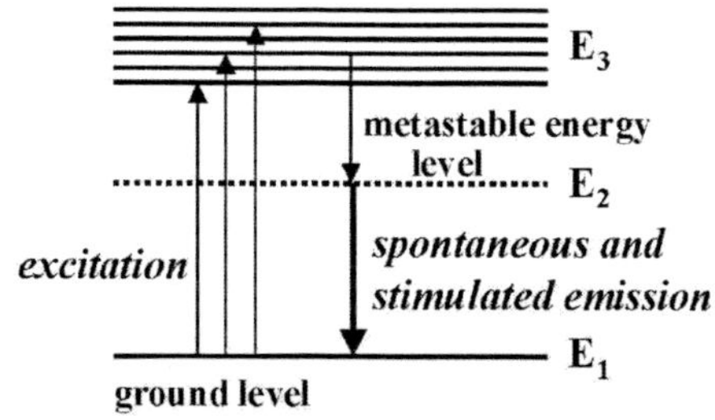

Fig. 13.11 Simplified energy-level diagram of a three-level laser.

The excited Cr^{3+}-ions can remain in the metastable energy level, E_2, for several milliseconds before spontaneous emission takes place. The first photons produced by the drop of excited electrons of the metastable energy level to the ground state, set off a "**chain reaction**" of stimulated emission, resulting in a high number of electrons making the same transition. This process produces a large number of photons that are in-phase and move in the same direction.

Since the ruby laser rod has a **totally reflective** mirror on one end and a **partially transmitting** mirror on the other side of the rod, an intense coherent laser beam is built up inside the rod and is then passed through the partially transmitting mirror. ***Note*** that this type of laser can only be operated in pulses (*pulsed laser*).

There are many **types of lasers** using gas, liquid, or solid media. The most important ones are listed below:

- **Al_2O_3:Cr** laser; ruby laser with a wavelength of 694.3 nm (see above)
- **Nd:YAG** laser; ~0.01% neodymium (Nd) is incorporated into the yttrium-aluminum-garnet (YAG) structure. This laser emits in the near-infrared at a wavelength of 1060 nm with a continuous power of about 250 watts (pulsed = several megawatts).
- **CO_2** laser; carbon dioxide lasers are some of the most powerful lasers made and operate mainly in the middle-infrared at 10.6 μm. The lasing process is initiated by *electron collisions* that excite *nitrogen molecules* to metastable energy levels, which subsequently transfer their energy and excite CO_2 molecules. The transition of the excited CO_2 molecules from the metastable energy level to the ground state produces a high-intensity photon beam.
- **semiconductor** lasers; semiconductor or *diode* lasers are the smallest in size (miniaturization). They consist of a *pn-junction* of a semiconducting compound such as GaAs that has a large-enough band gap to generate a laser beam. Inversion population is achieved by a strong forward bias of the heavily doped *pn*-junction, which generates a large number of *electron-hole pairs* that, upon **recombination**, emit light photons.

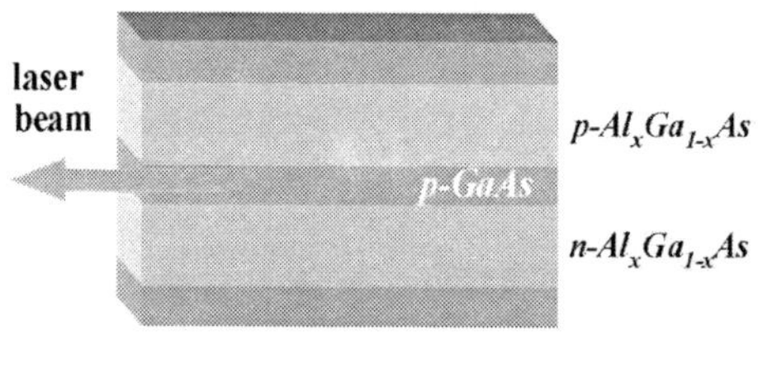

Fig. 13.12 Schematic of a semiconductor laser with a double heterojunction.

More recent developments use double heterojunctions to improve efficiency. Here, a thin layer of *p*-GaAs is confined between a layer of *p*- and *n*-$Al_xGa_{1-x}As$ (Figure 13.12). The AlGaAs layers have wider band gaps than the GaAs layer and lower refractive indices, which constrains the emitted laser light to proceed in a **miniature waveguide**.

13.6 Optical Fibers

Very thin optical SiO_2 fibers (~100 μm in diameter) are used for modern **communication systems**. These systems consist essentially of a *transmitter* (for example a semiconductor laser) to encode electrical signals to light signals, *optical fibers* to transmit light signals, and a *photodiode* to convert the light signals back into electrical signals.

Optical fibers for light transmission serve as **waveguides** for light signals. The retention of light within the optical fiber is made possible by having the light pass through a small central **core** of silica glass with a **higher refractive index** than the surrounding glass shell (see also total reflection). In this **single-mode** optical fiber type, only *one* guided light ray can be transmitted.

In a **multi-mode** optical glass fiber, *many* different rays can pass through the fiber simultaneously. This is accomplished by producing a fiber core with a **graded refractive index**, where SiO_2 is doped with GeO_2 or P_2O_5 to increase the refractive index.

Note that in order to use optical fibers for telecommunication systems, the *light loss* (attenuation) has to be extremely low (high purity fibers). The **attenuation** of an optical fiber is usually measured in *decibels per kilometer* (dB/km). Most new optical-fiber communication systems use *single-mode* fibers, because they have lower light losses and can be processed easier and more cheaply.

It is interesting to add a historical note: **A.G. Bell**[35] invented the **photophone** in 1878, using modulated sunlight falling on a photo-conducting selenium (Se) cell about 200 yards away.

Questions

What is the wavelength of ultraviolet and infrared light?

Which law describes the relation between refractive indices and refractive and incident angle?

How is phosphorescence defined?

What is the principle of cathode-luminescence?

What is the basic principle of a laser?

---

[35] **Alexander Graham Bell** (1847-1922), American scientist.

14 Electrical Properties

Similar to the way as we used *atomic bonding* for the classification of engineering materials, we will see that *electrical conductivity* is also related to the characteristics of atomic bonding, which is not surprising in light of the *electronic nature of atomic bonding.* In addition, this concept shows how strongly properties are related to the various crystal structures of materials.

14.1 Electrical Conduction

This section aims at the understanding how impurities, alloy additions, and temperature affect the electrical conductivity of materials (metals, semiconductors, ceramics, and polymers).

Electrical conductivity can, in general, be defined as a **movement of charge** from one location to another within a material. The different charge carriers can be:

- **electrons** (0.16×10^{-18} Coulomb)
- **holes** (absence of electrons)
- **cations**
- **anions.**

One simple method of measuring electrical conduction (current flow) is given in Figure 14.1. One of the *fundamental laws* of electrical conduction is **Ohm's law** [36], which describes the magnitude of **current flow**, I, through a circuit with a given resistance, R, and voltage, V:

$$V = I \cdot R \tag{14.1}$$

where the voltage is recorded in volts[37] [V], the current is measure in amperes[38] [A], 1A=1C/s, and the resistance is given in ohms [Ω].

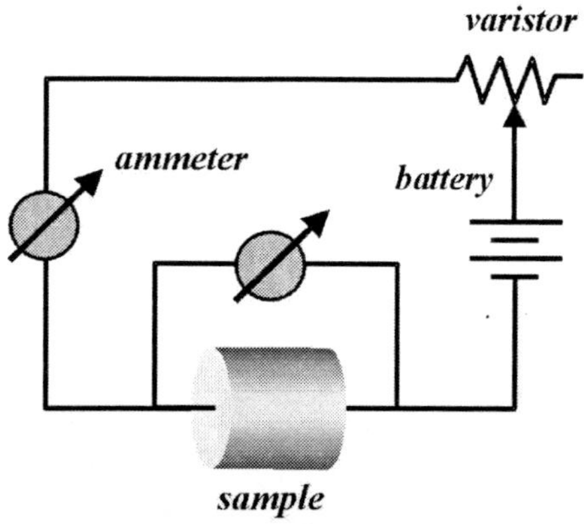

Fig. 14.1 Schematic of an electrical circuit to measure the electrical conductivity.

Note that R is the **resistance** of the electrical circuit (*circuit property*) and depends on *sample geometry.* Therefore, the *material property* **electrical resistivity,** ρ, is given in ohm·meter, [Ω m] and is defined as (Figure 14.2):

$$\rho = \frac{R \cdot A}{l} \tag{14.2}$$

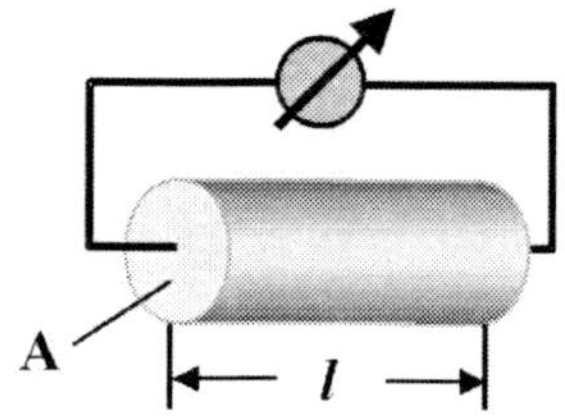

Fig. 14.2 Sample geometry (equation 14.2).

[36] **Georg Simon Ohm** (1787-1854), German physicist.

[37] **Alessandro Giuseppe Antonio Anastasio Volta** (1745-1827), Italian physicist.

[38] **André Marie Ampère** (1775-1836), French mathematician and physicist.

with l being the length of the wire and A the cross sectional area. ***Note*** that as l increases, R increases and as A increases, R decreases:

$$R = \frac{\rho \cdot l}{A} \qquad (14.3)$$

hence, there is only one way to actually change the resistivity (material property), ρ, by changing the material itself (its composition or structure).

Example 14.1

Calculate the required **length** of a metal wire to create a **resistance** of 200 Ω. Assume the diameter of the wire to be 0.01 cm and a conductivity of 5x10^5 (Ωm)$^{-1}$:

$$R = \frac{\rho \cdot l}{A},\ l = \frac{A \cdot R}{\rho},\ l = A \cdot R \cdot \sigma$$

$$l = \pi \cdot 0.005^2\, cm^2 \cdot 200\Omega \cdot 5 \cdot 10^3\, (\Omega cm)^{-1}$$

$$l = 78.5\, cm$$

The **electrical conductivity**, σ, (not a stress value) is defined as the passage or flow of current through a material instead of its resistance against the flow of current:

$$\sigma = \frac{1}{\rho} \qquad (14.4)$$

and is, therefore, given in (ohm·meter)$^{-1}$, [Ω m]$^{-1}$.

In general one can **categorize** materials regarding their specific electrical conductivities:

- **conductors:** $\mathbf{10^7\text{-}10^4}$ (Ωm)$^{-1}$
- **semiconductors:** $\mathbf{10^4\text{-}10^{-4}}$ (Ωm)$^{-1}$
- **insulators:** $\mathbf{10^{-4}\text{-}10^{-15}}$ (Ωm)$^{-1}$.

Pure metals (conductors) show the highest conductivities, while insulators (ceramics, glasses, and polymers) such as polyethylene and polystyrene reveal the lowest conductivities. ***Note*** that the electrical conductivity values range over *22 orders of magnitude*, as also shown in Table 14.1.

Electrical conductivity is the *product* of the **density** *of charge carriers*, n, the **charge** that is carried by each carrier, q, and the **mobility** of each carrier, μ, which leads to the fundamental equation:

$$\sigma = n \cdot q \cdot \mu \qquad (14.5)$$

where the units for n are [(m)$^{-3}$] for q are coulombs [C], and for μ are [m^2/(V·s)]. If either n, q, or μ increases, electrical conductivity also increases.

Tab. 14.1 Electrical conductivity of various materials.

Material Group	Material	Conductivity [Ω^{-1}m^{-1}]
Metals	Ag	6.3 x 10^7
	Cu	5.8 x 10^7
	Al	3.5 x 10^7
	Fe	10.3 x 10^6
	Steel	5.7-9.4 x 10^6
Semi-conductors	PbS	38.4
	Ge	2.0
	Si	4.0 x 10^{-4}
Ceramics Glasses	Al_2O_3	10^{-10} – 10^{-12}
	B-Si-glass	10^{-13}
	Nylon 66	10^{-12} – 10^{-13}
	C_2H_4	10^{-13} – 10^{-15}

Equation 14.5 can be used to explain **electrical behavior** of materials. For example in metals (conductors), the charge carriers are free electrons with a constant charge of minus one. The number of charge carriers (free electrons) is high and,

therefore, a large variation in charge mobility, μ, will control electrical conduction, σ.

The charge carrier mobility, μ, is the *average carrier velocity*, or the so-called **drift velocity**, $\bar{v}$, divided by the electrical field strength:

$$\mu = \frac{\bar{v}}{E} \qquad (14.6)$$

where $\bar{v}$ is given in [m/s] and is E has the units (volt/length) [V/m].
When both, *negative* and *positive* charge carriers contribute to the electrical conduction of a material, the above equation 14.5 has to be extended to:

$$\sigma = n_n q_n \mu_n + n_p q_p \mu_p \qquad (14.7)$$

where the subscripts n and p refer to negative and positive charge, respectively.

Note that for electrons, electron holes, and monovalent ions, the magnitude of q is $0.16 \text{x} 10^{-18}$ C. For multivalent ions, the charge q has to be multiplied by the valence of the ion, such as by 2 for oxygen, O^{2-}. Moreover note that the different charge carriers move in opposite directions.

Electrical conductivity is a function of *temperature*, *composition* and *deformation*; the effect that temperature change has on electrical conductivity of metals will be addressed first.

14.1.1 Temperature Effect

The overall effect on electrical conductivity observed in **metals** when the temperature rises is that conductivity decreases while resistivity increases. This behavior is due to the increase in **thermal vibration** with increasing temperature. The positive metal ion cores vibrate about their equilibrium position in the crystal lattice, creating thermally excited elastic waves (**phonons**). As a consequence, lattice disruptions are caused which reduce electron mobility, μ_e (due to the wave-like nature of electrons) and consequently increase resistivity:

$$\rho = \rho_{0^\circ C}(1+\alpha \cdot T) \qquad (14.8)$$

where $\rho_{0^\circ C}$ is the electrical resistivity at 0 °C and α is the temperature coefficient of resistivity (a positive number) with the units of [1/°C], and T stands for temperature.

It follows from equation 14.8 that **electrical resistivity of metals increases linearly with temperature**, as shown in Figure 14.3.

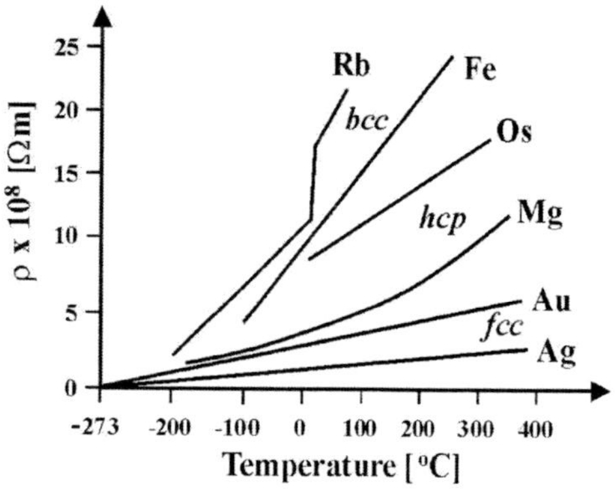

Fig. 14.3 Diagram of electrical resistivity of metals versus temperature. *Note* that resistivity increases with rising temperature.

14.1.2 Composition Effect

Similar to the effect observed in pure metals when the temperature rises, resistivity increases (at constant temperature) when **alloying elements** (impurities) are added to the system. While thermal vibration causes a lowering in electrical conductivity, due to a lowered electron mobility, solid-solution atoms also disrupt the "perfect" crystal structure and cause a **reduced electron mobility**, resulting in an increased resistivity of the alloy.

Figure 14.4 shows the electrical resistivity of copper doped with various alloying elements. Similarly, Figure 14.5 reveals a maximum in electrical resistivity for the Au-Cu-alloy composition of Au:Cu = 50:50.

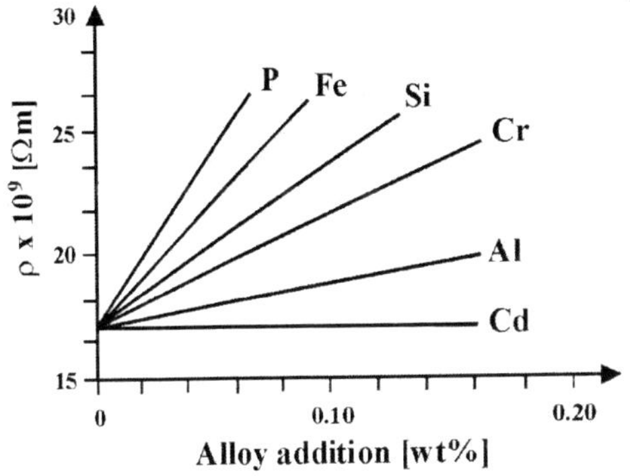

Fig. 14.4 Variation of electrical resistivity with composition of various copper alloys (T=const).

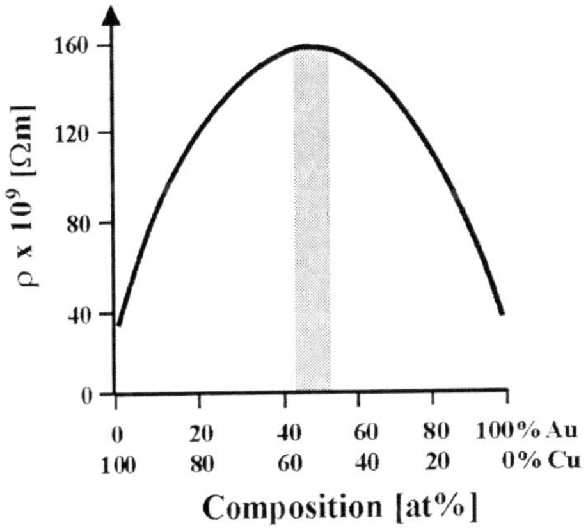

Fig. 14.5 Change in electrical resistivity with large compositional variations of Cu-Au alloys (T=const.).

14.1.3 Deformation Effect

As **deformation** of metals increases, the electrical conductivity, σ, decreases and resistivity increases. With a higher level of plastic deformation, a higher **dislocation density** forms within the material disrupting the crystal structure, which in turn **reduces electron mobility**, μ_e.

14.2 Energy Levels and Energy Bands

In order to understand the electrical properties of metals, semiconductors and insulators, one needs to understand the electron energy **band structure** of solids.

The ease or difficulty with which electrical conduction takes place in a material can be explained, when we return to the principle of **energy levels**. In solid materials, discrete energy levels give way to *energy bands*. The relative magnitude of the spacing between these bands determines the magnitude of electrical conduction.

A single atom shows *quantized* electron energy levels, which means that every electron is associated with a discrete energy state within the different **orbitals**; for example, as shown for chlorine:

Cl-orbitals: $1s^2, 2s^2, 2p^6, 3s^2, 3p^5$.

An uncharged (neutral) **Na^0 atom** has eleven electrons, two filling the 1s and 2s orbitals, six placed in the 2p orbital, and one residing in the 3s state. It follows that sodium reveals an electronic configuration of **$1s^2 2s^2 2p^6 3s^1$** (Figure 14.6). The outermost 3s-electron is a *valence electron* and can be involved in bonding with other atoms like Cl to form NaCl.

The distribution of electrons among the different energy states (orbitals) is given by the **Pauli exclusion principle**. This is an important concept in quantum mechanics, which says that no two electrons can occupy precisely the same energy level. Each orbital can be occupied by two electrons with *opposite* or **antiparallel electron spins**.

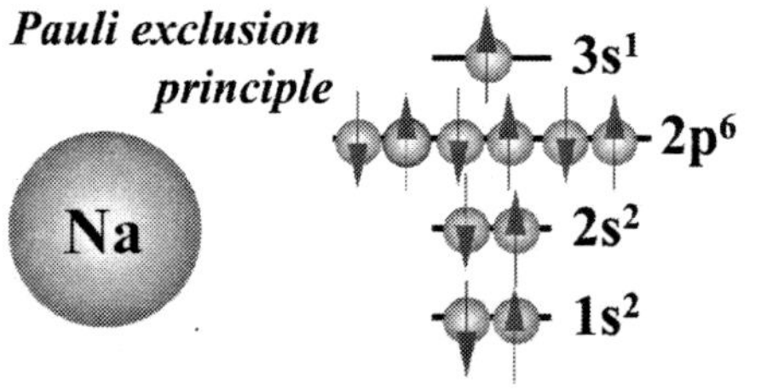

Fig. 14.6 Electronic structure of a non-charged sodium atom.

If we now consider a *hypothetical* four-atom molecule of sodium, $\mathbf{Na_4}$, the energy diagram of the core electrons, $1s^22s^22p^6$, stays essentially unchanged. However, the four *outer orbital electrons* are affected by the Pauli exclusion principle. Here, the **delocalized** outer electrons are **shared** by all four sodium atoms in the molecule. This delocalization (different energy levels) results in a **splitting** of the 3s energy level into four slightly different levels (Figure 14.7).

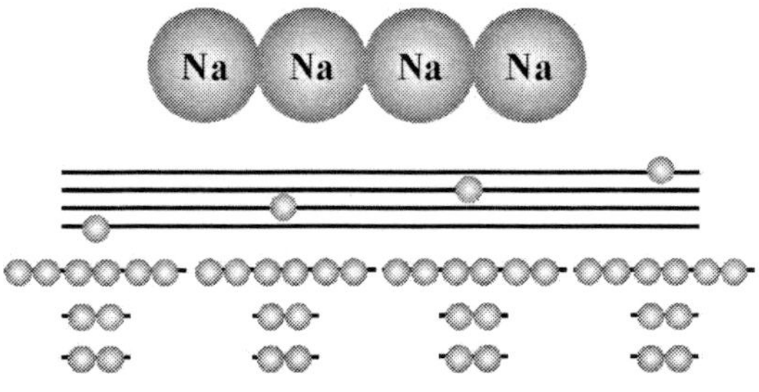

Fig. 14.7 Electronic configuration of a hypothetical Na_4-molecule. *Note* that the core electrons are not affected, but the four outer valence electrons are delocalized.

An important feature of the electronic configuration of the Na_4-molecule is that the 3s level is only **half filled.**

Note that the formation of **electron pairs** is *delayed* here. Four different energy levels rather than two with paired electrons are formed This phenomenon is referred to as **Hund's rule**[39]: Every orbital in a subshell is singly occupied with one electron before any of the orbital is doubly occupied. All electrons in *singly occupied* orbitals have the **same spin**.

If we extend the initial thought experiment to a large number of sodium atoms forming a **solid**, we see that for this metallic solid the *core electrons* again are essentially *not affected*. However, the large number of atoms involved produce a large number of 3s levels with slightly different energy (Figure 14.8).

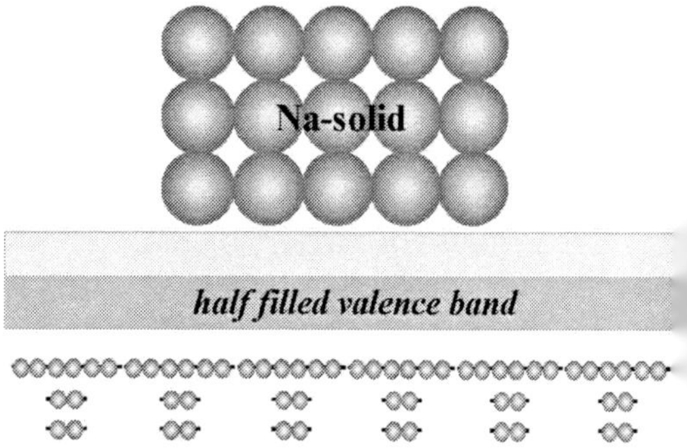

Fig. 14.8 Electronic structure of a Na-solid compound where half of the valence band is filled with 3s-electrons.

Since the spacing between these levels is extremely small, a *pseudo-continuous* **energy band** is formed. Since this band contains exclusively valence electrons, it is

[39] **Friedrich Hermann Hund** (1896-1997), German physicist.

termed the **valence band**. Moreover, because this band is only half-filled, the electrons can move more freely, and reveal high mobility.

An important conclusion from the above concept is that **metals** are *good electrical conductors* because their *valence band is only partially filled* and their electrons are *highly mobile*.

14.3 Fermi Level

A more detailed picture of the nature of electrical conduction in metals is gained by considering how the energy band changes with temperature.

In the case of the sodium metal it was implied that the valence band was only filled up to the half point of the band and is completely empty above. This is only a true picture at a temperature of absolute zero (0 K). At *zero Kelvin*, the energy level of the **highest filled state** is called the **Fermi-energy**[40], E_f, as shown in Figure 14.9.

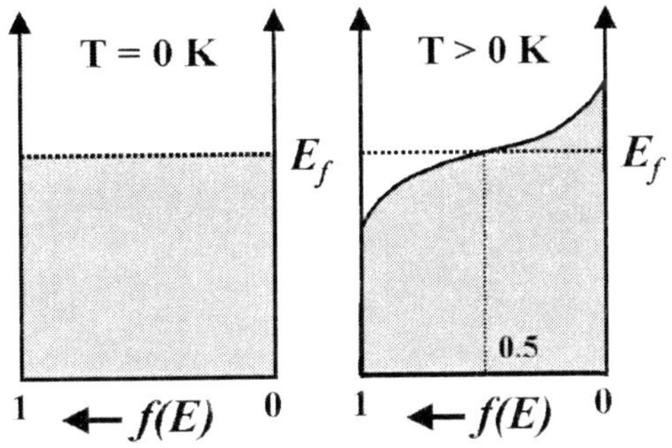

Fig. 14.9 The Fermi function describes the relative filling of an energy level depending on temperature.

[40] **Enrico Fermi** (1901-1954), Italian physicist.

The extent to which a given energy level is filled depends on the respective temperature and is described by the **Fermi-function** (probability function):

$$f(E) = \frac{1}{e^{(E-E_f)/kT} + 1} \qquad (14.9)$$

where k is the Boltzmann constant, T is the temperature in Kelvin, and E_f is the Fermi energy. The Fermi function, $f(E)$, represents the *probability* that an energy level, E, is occupied by an electron and can have values between 0 and 1. At 0 K, the Fermi function is 1 up to the Fermi level (Fermi energy) and equal to 0 above E_F. ***Note*** that this limiting case (0 K) consequently means that the material is **not electrically conductive** (Figure 14.10).

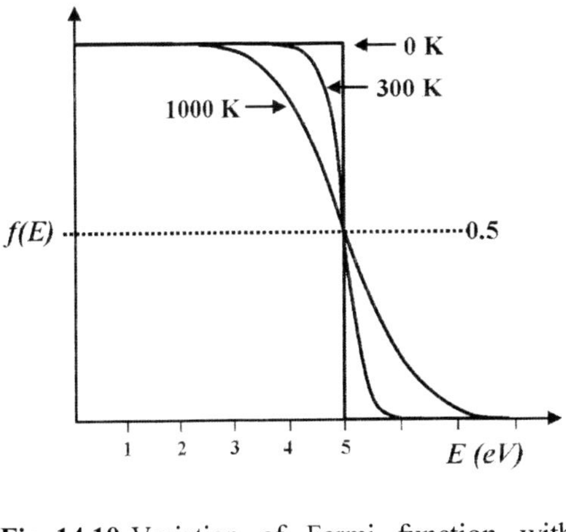

Fig. 14.10 Variation of Fermi function with temperature for a metal with a Fermi level of 5.0 eV.

The Fermi function for T = 0 K is a **step function**; however, at higher temperatures (T > 0 K), the Fermi function varies between 1 and 0 in a smooth fashion and the Fermi energy is precisely 0.5.

The range over which the Fermi function drops from 1 to 0 increases with increasing temperature and is on the order of kT.
One important requirement for electrical conduction is therefore that some **unfilled energy states** (energy levels) exist so that electrons can have energies *above* the Fermi energy level, E_f.

Since all energy levels below E_F are filled, conduction requires electrons to **increase** their **energy** to some level *above* the Fermi energy. One way to obtain this energy promotion is via *thermal energy input.* Hence, heating a metal to room temperature promotes electrons into the unfilled energy levels above E_f, thus contributing to electrical conduction.

As a consequence, **metals** are **good conductors** because thermal energy is sufficient (even at RT) to promote electrons above the Fermi energy level to unoccupied states. The accessibility of unoccupied energy levels in adjacent atoms yields a high mobility of the conduction electrons (free electrons) and results in the good electrical conductivity known for metals.

One important application of conducting metals is the measurement of temperature via a **thermocouple**, which involves two metal wires. If a metal wire is connected between two different temperatures (hot and cold), more electrons are excited at the hot end of the wire, which creates a driving force for electron transport from the hot to the cold zone, which is then negatively charged with a voltage V_a.

V_a only depends on the temperature difference $(T_1 - T_2)$ but not on the temperature gradient within the wire. If a second metal wire is connected, as shown in Figure 14.11, a voltmeter will measure the net voltage $V_{net} = V_a - V_b$, which is very sensitive to small changes in temperature.

This effect was named the **Seebeck effect** [41] after its discoverer and the induced (net) voltage is termed Seebeck potential.

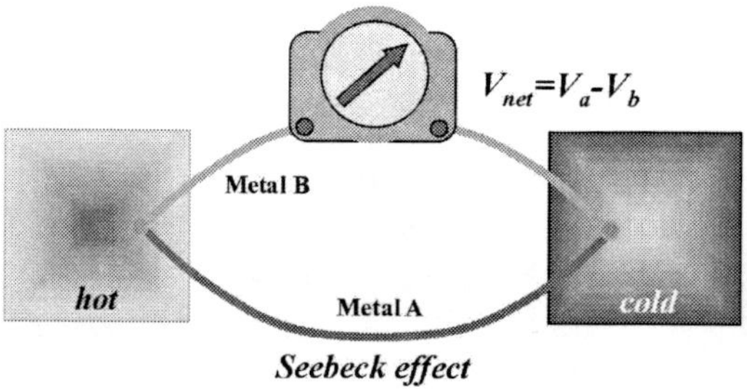

Fig. 14.11 Schematic illustrating the Seebeck effect, which allows the detection of small temperature differences.

Note that the basic principle of the Seebeck effect is related to the temperature-sensitivity of the Fermi function, which is different for different metals.

14.4 Insulators

In the case of a **non-metallic** solid such as carbon in the *diamond structure*, there is very poor electrical conductivity. The valence electrons in this covalently bonded solid are **shared** among adjacent atoms. Consequently, the *valence band* of diamond is *completely filled.*

---

[41] **Thomas Johann Seebeck** (1170-1831), Russian-German physicist.

This **valence band** corresponds to the *sp^3-hybrid energy level* of an isolated carbon atom. To promote electrons to energy levels above this sp^3 level in an isolated carbon atom requires going above regions of forbidden energy.

Similarly, the promotion of an electron above the sp^3-energy level of the valence band into the conduction band within a solid requires going above an **energy band gap.** Electrons are forbidden to have energies within the band gap.

The concept of the **Fermi energy** level, E_F, still applies. However, here the Fermi energy level falls right into the center of the energy band gap, E_g.

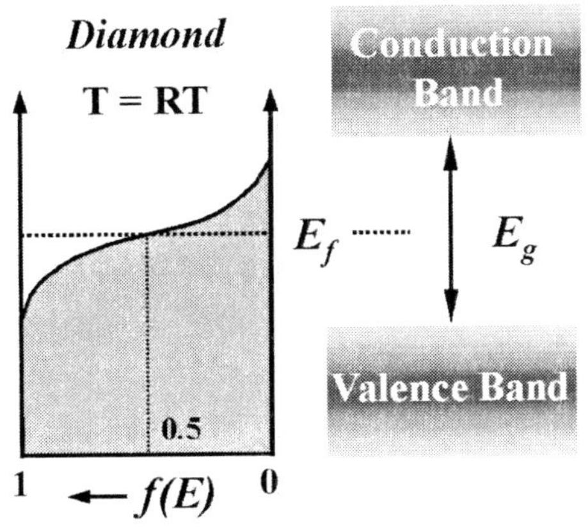

Fig. 14.12 Comparison of Fermi function with energy band structure of a wide band gap material (diamond; E_g = 6.0 eV).

Tab. 14.2 Band gap values (in [eV]) of various insulating materials.

Wide Band Gap Materials			
Fe_2O_3	3.100	NaF	6.667
ZnO	3.200	KCl	7.000
AsCl	3.200	NaCl	7.300
TiO_2	3.400	MgO	7.800
ZnS	3.600	Al_2O_3	8.000
CoO	4.000	**SiO_2**	**8.000**
NiO	4.200	BaF_2	8.857
PbF_2	4.276	Al_2O_3	8.857
Ga_2O_3	4.600	SrF_2	9.538
BN	4.800	MgF_2	10.973
UO_2	5.200	LiF	12.000
Diamond	6.000	CaF_2	12.000
CdF_2	6.200	MnF_2	15.500

Note that the Fermi function is basically 1 within the valence band and 0 throughout the conduction band, as schematically shown in Figure 14.12.

14.5 Semiconductors

Comparing the behavior of **diamond** with **silicon**, which has the same crystal structure, shows that the Fermi function of silicon has features very similar to diamond (Figure 14.13).

The major difference, however, is that silicon has a **smaller band gap** than diamond (compare Tables 14.2 and 14.3). The band gap, E_g, of diamond and silicon are 6.0 eV and 1.1 eV, respectively. Therefore, in Si a small but considerable number of electrons are **promoted** by thermal energy from the valence band into the conduction band at room temperature.

Note that since **thermal energy** is sufficient (even at room temperature) to promote electrons, **electron holes** are continuously produced within the valence band and, moreover, that the number of holes *equals* the number of conduction electrons.

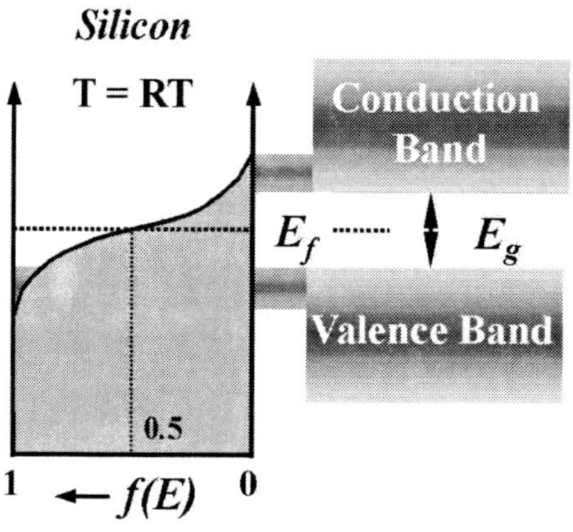

Fig. 14.13 Comparison of Fermi function with energy band structure of an intrinsic semiconductor (silicon; E_g = 1.1 eV).

These electron **holes** are **positive charge carriers**. With both positive and negative charge carriers only in moderate number, however, pure silicon reveals a moderate electrical conduction at room temperature.

Tab. 14.3 Band gap values (in [eV]) of different semiconducting materials.

Band Gap of Semicoductors			
KBr	0.185	CdO	2.100
PbTe	0.275	GaP	2.250
PbS	0.350	CdS	2.420
PbSe	0.400	TlBr	2.480
Ge	0.740	ZnSe	2.600
Si	**1.120**	$BaTiO_3$	2.800
GaAs	1.400	AsI	2.800
CdTe	1.450	AgBr	2.800
CdSe	1.850	α-SiC	2.900
Cu_2O	2.100		

Such pure undoped semiconductors are called **intrinsic semiconductors**. In these materials, holes and electrons are produced in *pairs* and thus the number of conduction electrons equals the number of holes, so that it follows:

$$n_i = e = h \qquad (14.10)$$

where n_i is the intrinsic charge carrier concentration (carriers per unit volume).

This means that our initial equation on electrical conduction, $\sigma = nq\mu$, becomes:

$$\sigma = n_i \cdot q \cdot (\mu_e + \mu_h) \qquad (14.11)$$

Note that the mobility of electrons is always greater than that of holes. For intrinsic silicon, the electron mobility (0.135 $m^2/(Vs)$) is 2.81 times greater than the corresponding mobility of holes (0.048 $m^2/(Vs)$ at 300 K). The ratio of electron-to-hole mobility for intrinsic germanium is 2.05 at 300 K.

14.5.1 Temperature Effect

The effect of temperature on the electrical conductivity of intrinsic semiconductors can be described as follows: **at 0 K** the valence band is completely filled and the conduction band is completely empty, which means there is no electrical conduction at 0 K (similar to metals).

At temperatures slightly **above 0 K**, some of the valence electrons are **thermally activated** and excited across the energy gap into the conduction band, creating *electron-hole pairs*. Therefore, in contrast to metals whose conductivities decrease with increasing temperature (thermal vibration - lattice disruption), the conductivity of semiconductors **increases** with increasing temperature.

Due to this activation process, there is a **temperature dependence**, similar to many thermally activated processes.
The concentration of electrons with sufficiently high thermal energy to enter the conduction band, n_i, varies according to:

$$n_i \propto e^{-(E_g - E_{av})/kT} \qquad (14.12)$$

where E_g is the energy band gap, E_{av} is the average energy across the band gap, k stands for the Boltzmann constant and T is the absolute temperature.

For intrinsic semiconductors, E_{av} is half-way across the band gap, which equals ½ E_g. Thus equation 14.12 becomes:

$$n_i \propto e^{-(E_g - E_g/2)/kT} \qquad (14.13)$$

$$n_i \propto e^{-E_g/2kT} \qquad (14.14)$$

Since the electrical conductivity of intrinsic semiconductors is **proportional** to the concentration of electrical **charge carriers**, n_i, the above equation can be expressed as:

$$\sigma = \sigma_0 \cdot e^{-E_g/2kT} \qquad (14.15)$$

and in a natural logarithmic form as:

$$\ln \sigma = \ln \sigma_0 - \frac{E_g}{2kT} \qquad (14.16)$$

where σ_0 is an overall constant that primarily depends on the mobilities of electrons and holes. The small temperature dependence of σ_0 is neglected here.

Note that since equation 14.16 represents an *Arrhenius-like* correlation, E_g can be determined from the slope of a diagram of $\ln \sigma$ versus $1/T$ (in K), as shown in Figure 14.14.

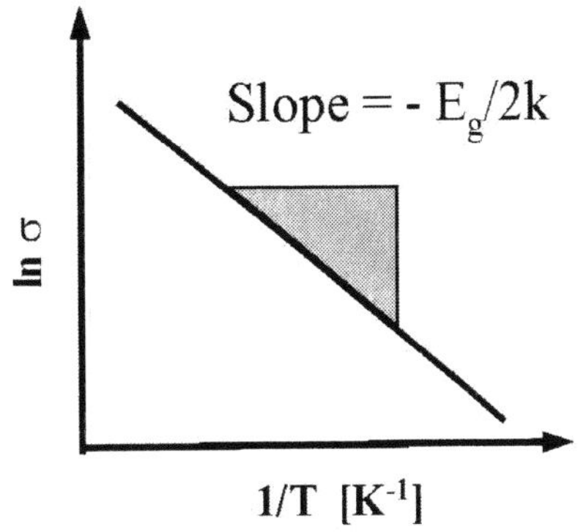

Fig. 14.14 Diagram of ln conductivity versus 1/T, which allows the determination the band gap energy, E_g.

14.5.2 Extrinsic Semiconductors

Extrinsic semiconductors are dilute **substitutional solid solutions** in which the solute impurity atoms have different valence characteristics from the solvent atomic lattice. The concentrations of the added impurity atoms in these semiconductors are usually on the order of 100 to 1000 parts per million (ppm).

14.5.2.1 *n*-type

Consider a two-dimensional covalent bonding model of a pure (undoped) silicon crystal, as shown in Figure 14.15.

If an **impurity** atom of group VA element, such as phosphorous, replaces a silicon atom, which is a group IVA element, there will be one **excess electron** in addition to the four electrons needed for the tetrahedral covalent bonding in the silicon lattice (see Figure 14.16).

This extra electron is only **loosely bonded** to the positively charged phosphorous nucleus and has a binding energy of only **0.044 eV** (at RT). This energy is about 4 % of the energy required for a valance electron to jump the band gap (1.12 eV) into the conduction band. Therefore, only 0.044 eV of energy is required to **remove** the excess electron from its parent phosphorous nucleus so that it can participate in electrical conduction.

Fig. 14.15 Two-dimensional bonding model of pure silicon.

If an **electric field** is externally applied, this extra electron becomes a **free electron** available for conduction and the remaining **phosphorous** atom becomes **ionized** and acquires a positive charge, +1, as illustrated in Figure 14.17.

Group VA elements such as **P**, **As**, and **Sb**, when added to intrinsic semiconductors such as Si or Ge, provide easily ionized impurity atoms and free electrons which contribute to electrical conduction.

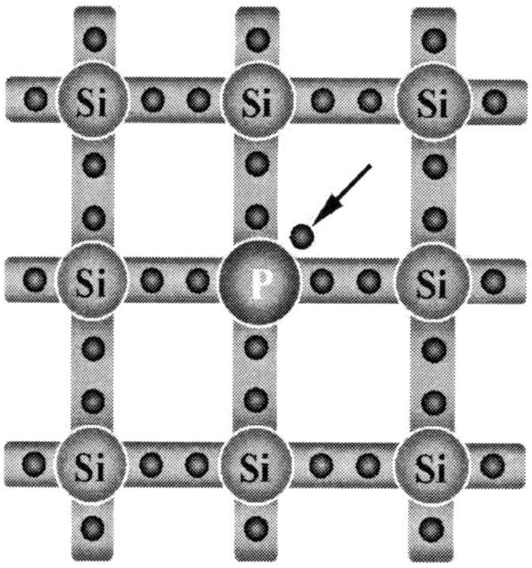

Fig. 14.16 Two-dimensional bonding model of P-doped silicon.

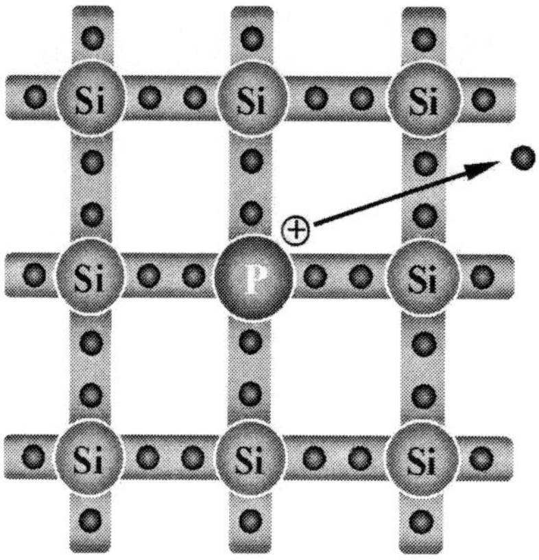

Fig. 14.17 Two-dimensional bonding model of P-doped silicon under the influence of an electric field.

Since these group VA elements donate conduction electrons, they are called **donor impurity atoms**. Si or Ge doped with VA group elements are therefore called ***n*-type** (*n* for negative) **extrinsic semiconductors**.

Note that in terms of the energy-band diagram, the extra electron of a group VA element occupies an energy level in the forbidden energy gap just slightly below the empty conduction band (Figure 14.18).

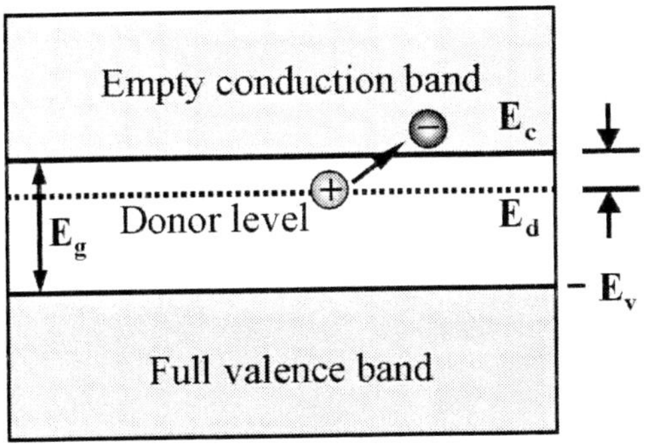

$$\Delta E = E_c - E_d$$

Fig. 14.18 Energy-band diagram of an *n*-type extrinsic semiconductor (doped with phosphorous).

Such an energy level is called a **donor level**, since it is provided by the donor impurity atom. When the extra electron is excited into the conduction band, an immobile positive ion, P^+, is left behind.

14.5.2.2 *p*-type

When a trivalent group IIIA element such as boron, B, is substitutionally introduced in the tetrahedrally bonded Si lattice, one of the bonding orbitals is actually missing and a **hole is created** in the bonding structure of silicon, as shown in Figure 14.19.

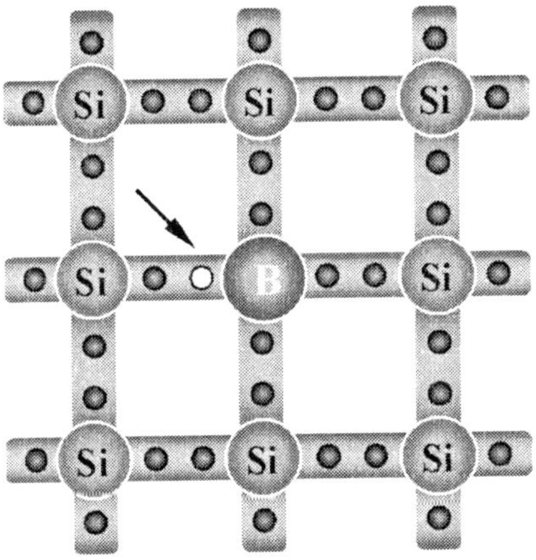

Fig. 14.19 Two-dimensional bonding model of B-doped silicon.

If an **external electric field** is applied, one of the neighboring electrons from another tetrahedral Si bond can attain sufficient energy to break loose from its bond and move to the missing bond (hole) of the impurity atom (boron), as illustrated in Figure 14.20.

When the hole associated with the boron atom is filled by an electron of a neighboring silicon atom, boron becomes **ionized** and acquires a negative charge of −1. The binding energy associated with the **removal** of the electron from a neighboring silicon atom, thereby creating an electron hole, and the **subsequent transfer** of the electron to the boron atom requires only **0.045 eV**.

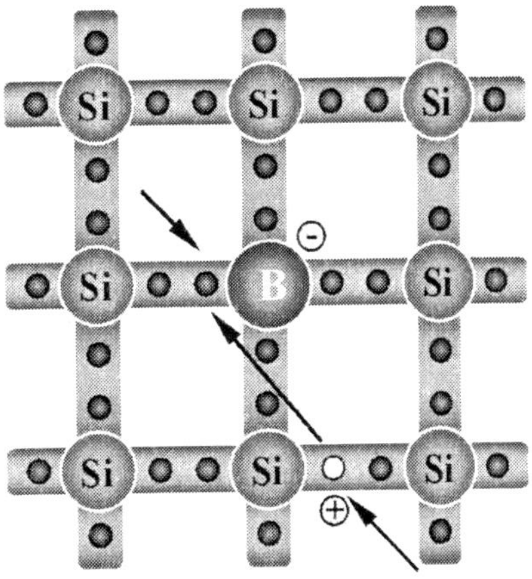

Fig. 14.20 Two-dimensional bonding model of B-doped Si with electron transfer and hole formation.

Again, the energy required to **remove** *and* **transfer** the electron is small compared to the energy of the band gap of silicon (1.1 eV). In the presence of an applied electric field, the hole created by ionization of a boron impurity atom behaves as a positive charge carrier and migrates in the silicon toward the negative terminal.

In terms of the energy-band diagram, the boron atom provides an energy level called the **acceptor level**, which is slightly higher (0.045 eV) than the uppermost level of the full valence band, as shown in Figure 14.21.

Group IIIA elements such as **B**, **Al**, and **Ga**, which provide acceptor levels in the semiconductors, are called *acceptor atoms*. Since the majority of charge carriers in these extrinsic semiconductors are **holes** in the valence bond structure, they are called ***p*-type extrinsic semiconductors**.

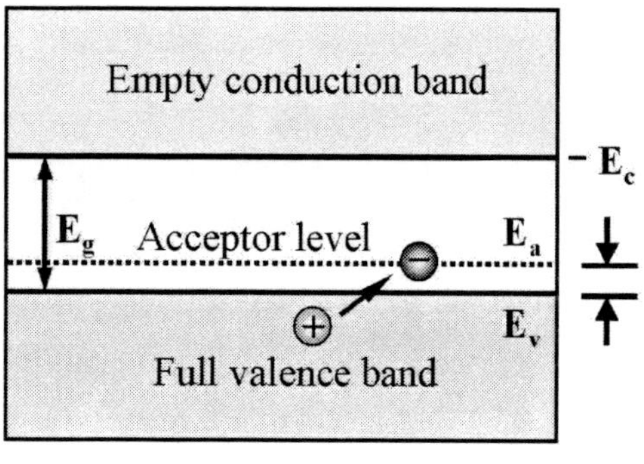

$$\Delta E = E_a - E_v$$

Fig. 14.21 Energy-band diagram of an *p*-type extrinsic semiconductor (doped with boron).

The process of adding small amounts of substitutional impurity atoms to silicon to generate an extrinsic semiconducting material is called **doping**. Figure 14.22 shows the effect of various dopants on the ionization energies.

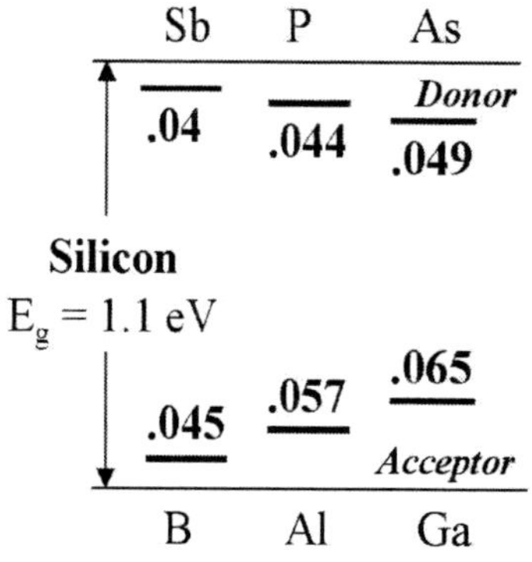

Fig. 14.22 Ionization energies of different acceptor and donor elements in silicon (given in [eV]).

14.5.2.3 Charge Densities

In semiconductors such as Si and Ge, mobile electrons and holes are constantly **generated** and **recombined**. At constant temperature and under equilibrium conditions, the product of the negative free electrons and positive holes concentrations is a **constant**,

$$n \cdot p = n_i^2 \qquad (14.17)$$

where n_i is the intrinsic concentration of charge carriers which is a constant for a given temperature. This equation is valid for intrinsic and extrinsic semiconductors. In an extrinsic semiconductor, the increase in one type of carrier (*n* or *p*) reduces the concentration of the other through recombination.

The type of charge carrier with the higher concentration in extrinsic semiconductors is designated the **majority carrier** and the carrier type with the smaller concentration is termed the **minority carrier**, as shown in Table 14.4.

Tab. 14.4 Majority versus minority charge carrier concentrations in extrinsic semiconductors.

Carrier Concentration		
Semi-conductor	**Majority** Carrier Concentration	**Minority** Carrier Concentration
n-type	n_n - **electron** concentration in *n*-type sc.	p_n - **holes** concentration in *n*-type sc.
p-type	p_p - **hole** concentration in *p*-type sc.	n_p - **electron** concentration in *p*-type sc.

The concentration of electrons in an *n*-type semiconductor is commonly denoted by n_n and that of holes in a *p*-type material by p_n. Similarly, the concentration of holes in a *p*-type semiconductor is given by p_p and that of electrons by n_p (*p*-type).

The second fundamental relationship of extrinsic semiconductors is that they have to be **charge neutral**. The total crystal must therefore be electrically neutral, which means that the charge density in each volume must be zero.

There are *two types* of charged particles in extrinsic semiconductors:

- immobile ions (donor or acceptor ions)
- mobile charge carriers (electrons and holes).

Note that the total negative charge density must equal the total positive charge density. The total negative charge density equals the sum of negative acceptor ions, N_a, and the electrons, n, $(N_a + n)$, while the total positive charge density equals the sum of the positive donor ions, N_d, and the holes, p, $(N_d + p)$:

$$N_a + n = N_d + p \quad (14.18)$$

In an ***n*-type semiconductor** (created by adding donor impurity atoms) $N_a = 0$. Since the number of electrons is much higher than the number of holes and the above equation reduces to:

$$n_n \approx N_d \quad (14.19)$$

Thus, in an *n*-type semiconductor the free-electron concentration is approximately equal to the concentration of donor atoms. The corresponding concentration of **holes** can be obtained by (with $n_n \cdot p_n = n_i^2$):

$$p_n = \frac{n_i^2}{n_n} \approx \frac{n_i^2}{N_d} \quad (14.20)$$

and consequently, for a ***p*-type semiconductor** (with $p_p \approx N_a$) it follows:

$$n_p = \frac{n_i^2}{p_p} \approx \frac{n_i^2}{N_a} \quad (14.21)$$

A **typical concentration** for **intrinsic** silicon at 300 K is $n_i = 1.5 \cdot 10^{16}$ carriers/m^3 and for **extrinsic** Si (doped, for example, with arsenic) is $n_i = 10^{21}$ electrons/m^3.

Example 14.2

Calculate the majority and minority carrier concentration for silicon doped with 10^{21} phosphorous atoms/m^3.

Since silicon is doped with phosphorous, the extrinsic silicon is *n*-type and it follows: **electrons** per m^3:

$$n_n = N_d = 10^{21}$$

holes per m^3:

$$p_n = \frac{n_i^2}{N_d} = \frac{(1.5x10^{16} \cdot m^{-3})^2}{10^{21} \cdot m^{-3}} = 2.25 \cdot 10^{11}$$

Example 14.3

The electrical resistivity of pure silicon is 2.3·10^3 Ωm at room temperature (300 K). Calculate the electrical resistivity of Si at 473 K. Use $E_g = 1.1$ eV and $k = 8.62 \cdot 10^{-5}$ eV/K.

$$\sigma = \sigma_0 \cdot e^{-E_g/2kT}, \quad \sigma_{473} = \sigma_0 \cdot e^{-E_g/2kT_{473}}$$

$$\sigma_{300} = \sigma_0 \cdot e^{-E_g/2kT_{300}}$$

$$\frac{\sigma_{473}}{\sigma_{300}} = \exp\left(\frac{-E_g}{2kT_{473}} + \frac{E_g}{2kT_{300}}\right)$$

$$\frac{\sigma_{473}}{\sigma_{300}} = \exp\left(\frac{-1.1eV}{2(8.62 \cdot 10^{-5}\, eV/K}\left(\frac{1}{473\,K} - \frac{1}{300\,K}\right)\right)$$

It results a Si-conductivity of 1.04 (Ωm)$^{-1}$ at 473 K, which is about a factor of 2400x higher than at room temperature.

14.5.2.4 Temperature Effect

The electrical conductivity of an **extrinsic** semiconductor such as silicon, which contains doped impurity atoms, is affected by temperature. At lower temperatures, the number of **impurity atoms** per unit volume that are **activated** (ionized) determines the electrical conductivity of the silicon. As the temperature increases, more and more of the impurity atoms are ionized and thus the electrical conductivity of extrinsic silicon **increases** with increasing temperature (in the extrinsic range), as shown in Figure 14.23.

In the **extrinsic range**, only a small amount of energy is needed to ionize the impurity atoms. The amount of energy needed to excite the donor electron (*n*-type silicon) into the conduction band is $(E_c - E_d) =$ 0.044 eV (for phosphorous). Hence, the slope of the diagram *ln σ* versus *1/T* in the extrinsic region is:

$$slope = -(E_c - E_d)/k \qquad (14.22)$$

For a certain temperature range above which **complete** ionization of the donor atoms occurs, an increase in temperature does not substantially change the electrical conductivity of the *n*-type extrinsic semiconductor.

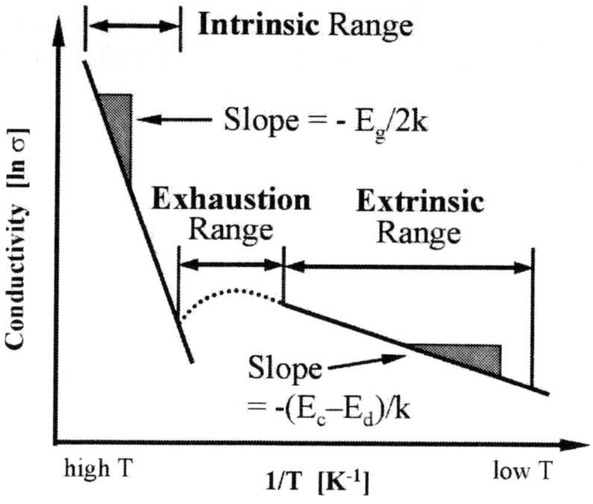

Fig. 14.23 Temperature dependence of the electrical conductivity of extrinsic semiconductors (plot of lnσ vs. 1/T).

For an ***n*-type** semiconductor, this temperature range is termed the **exhaustion range**, since donor atoms become completely "exhausted" (ionized) after the loss of their donor electrons.

For ***p*-type** semiconductors, this temperature range is referred to as the **saturation range** since acceptor atoms become completely "saturated" (ionized) with acceptor electrons (from the surrounding silicon).

As the **temperature increases** beyond that of the exhaustion or saturation range, the intrinsic range is entered upon. The higher temperatures provide sufficient **activation energies** for electrons to jump the semiconductors band gap, E_g, so that **intrinsic conduction** then becomes **dominant** (see Figure 14.23).

14.6 Semiconductor Devices

Most common semiconductor devices depend on the properties of the boundary between *p*-type and *n*-type materials, so-called ***pn*-junctions**.

14.6.1 *pn*-Junction Diode

A *pn*-junction diode can, for example, be produced by growing a single crystal of intrinsic silicon and first doping it with *n*-type material and then with *p*-type material. More commonly, however, a *pn*-junction is produced by solid-state diffusion of one type of impurity such as *p*-type into existing *n*-type material.

Before joining a *p*-type and *n*-type semiconductor material, both segments are **electrically neutral**. The majority carriers are holes in the *p*-type segment, while electrons are the majority carriers in the *n*-type segment of the diode. Upon joining the two segments, majority carriers at and near the *pn*-junction will counter-diffuse across the interface and **recombine**. Since the remaining ions at the interface region are larger (and heavier) than electrons and holes, they will reside at their position.

After several recombinations of majority carriers at and near the interface, this process will come to an end, since the electrons crossing the interface are then **repelled** by the negative charge of the ions. Similarly, holes crossing the interface will be repelled by the positive charge of the ions in the *n*-type material.

The immobile ions at the interface region create a zone that is depleted of majority carriers (electrons and holes) and is referred to as **depleted zone**, as shown in Figure 14.24.

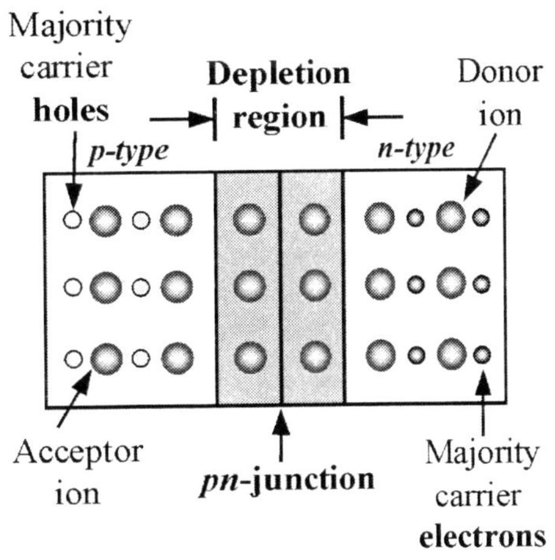

Fig. 14.24 Formation of a depleted zone at the interface region of a *pn*-junction (loss of majority carriers).

Note that under open circuit conditions there exists a positive potential difference or a barrier to majority carrier flow. Therefore, there is **no net current flow** under open circuit conditions.

However, when an external voltage is applied to a *pn*-junction, this junction becomes **biased**. If the *n*-type material of the *pn*-junction is connected to the positive terminal of a battery and the *p*-type material is connected to the negative terminal, the diode is **reverse biased** (Figure 14.25).

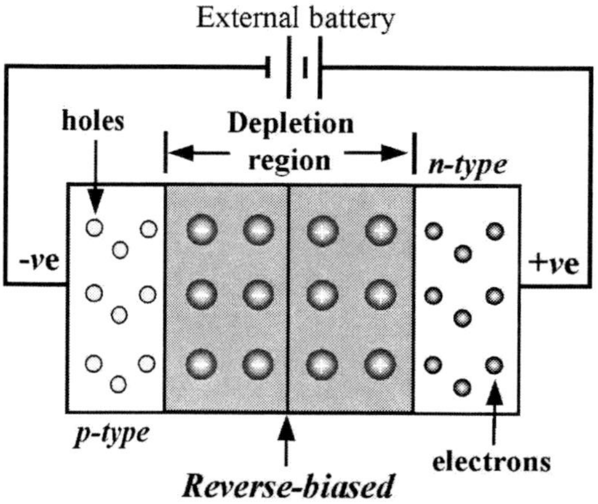

Fig. 14.25 Reverse-biased *pn*-junction diode where the majority carriers are attracted away from the interface region.

With this type of arrangement, the majority carriers of the *n*-type material (electrons) are attracted to the positive terminal of the battery, while the majority carriers of the *p*-type material (holes) are attracted to the negative terminal. Consequently, the **depletion region is widened** under reverse-biased condition and there is no current flow of majority carriers.

Note that under reverse-bias conditions thermally generated **minority carriers** (holes in the *n*-type material and electrons in the *p*-type material) are driven toward the *pn*-junction where they recombine. Hence, a small current flow of minority carriers is monitored under a reverse-bias setup, which is termed **leakage current**.

In contrast, if the *n*-type material of the *pn*-junction diode is connected to the negative terminal of an external battery and the *p*-type material is connected to the positive terminal, the diode is said to be **forward-biased.**

In this configuration, electrons are repelled from the negative terminal of the battery toward the *pn*-junction and holes are correspondingly repelled away from the positive terminal, as schematically shown in Figure 14.26.

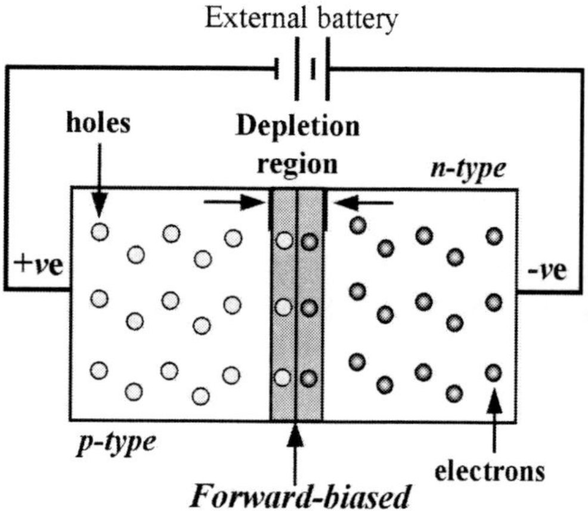

Fig. 14.26 Forward-biased *pn*-junction diode where the majority carriers are repelled toward the interface region lowering the energy barrier for current flow.

Under forward-bias condition, the **energy barrier** of the *pn*-junction is greatly *reduced* so that some electrons and holes are able to **cross the junction** and recombine.

During the forward bias of a *pn*-junction, **electrons** from the battery **enter** the *n*-type material of the diode. For every electron that crosses the junction and recombines with a hole, another electron enters from the negative terminal of the battery. Similarly, for every hole that recombines with an electron in the n-type material, a **new hole is formed** when an electron leaves the p-type material and flows toward the positive terminal of the battery. Since the energy barrier is greatly reduced for majority carrier flow under forward bias, a considerable current is measured, as shown in Figure 14.27.

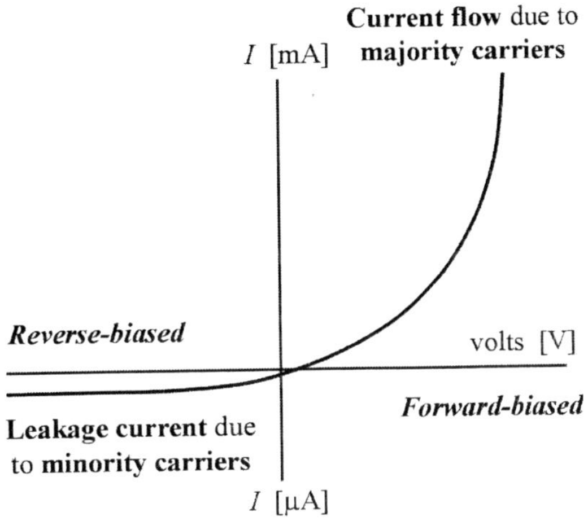

Fig. 14.27 Current-voltage characteristic of a pn-junction diode.

One important **application** of pn-junction diodes is to convert alternating current into direct current, which is termed **rectification**. Thus, diodes used for this particular process are commonly called *rectifier diodes*.

When an **ac-signal** is applied to a pn-junction diode, the diode will only conduct current when a positive voltage is applied to the p-type material (forward-biased; compare Figures 14.26 and 14.27). As a result, a half-wave rectification will be monitored, as illustrated in Figure 14.28.

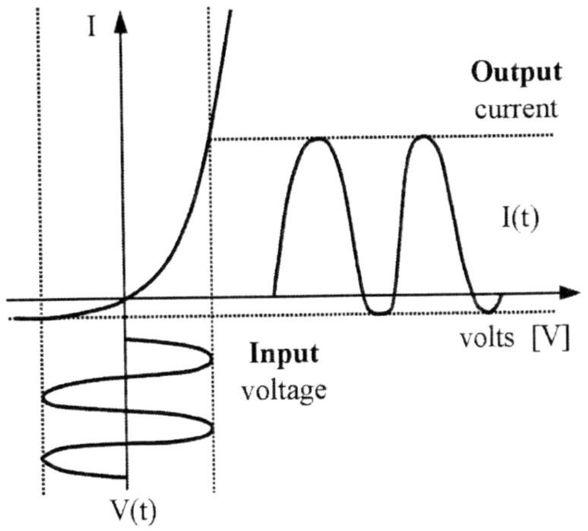

Fig. 14.28 Current-voltage characteristic of a rectifying diode converting ac-current into a dc-signal.

Note that this output signal can be further smoothed with other electronic devices and circuits so that a steady **dc-signal** is formed.

14.6.2 *npn*-Junction Transistor

A bipolar-junction transistor (BJT) is an electronic device that servers as a current **amplifier**. Such a device consists of a pair of np-, or pn-junctions, depending whether electrons (np) or holes (pn) are the overall current source. Figure 14.29 shows a schematic of a npn-bipolar junction transistor.

The three different regions of the transistor are termed **emitter**, **base**, and **collector**. The left np-junction (emitter) is forward-biased and hence the emitter is n-type, electrons are emitted.

The **base** of the transistor controls the flow of charge carriers and is p-type for a npn-transistor (and n-type for a pnp-transistor). The base is commonly processed very thin, less than 1 μm, and **lightly doped** so that only a small fraction of the electrons from the emitter will recombine with the oppositely charged majority charge carriers (holes) of the base.

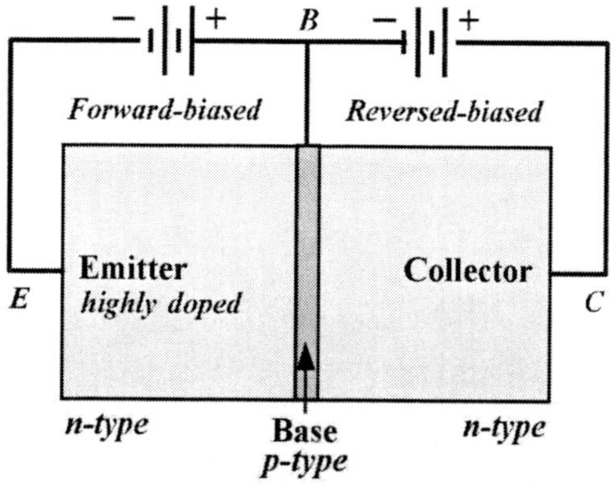

Fig. 14.29 Schematic of a *npn*-bipolar junction transistor with an emitter, a base, and a collector region.

The **collector** of the BJT mainly collects the electrons from the emitter. Once the electrons have passed the base regions and entered the collector, they can move freely as majority charge carriers.

Under normal operating conditions of an *npn*-bipolar junction transistor, the emitter-base junction is **forward-biased**, while the collector-base junction is **reverse-biased** (Figure 14.29). The forward bias of the emitter-base junction causes an injection of electrons from the emitter into the base. Although some of the electrons recombine in the base region, the majority of electrons pass through the thin base region into the collector, where they are attracted by the positive terminal of the collector.

Since the electron flow across the base region into the collector (overshoot) is an **exponential function** of the emitter voltage, V_e, the current in the collector, I_c, is an exponential function of V_e:

$$I_c = I_0 e^{V_e/B} \tag{14.23}$$

where I_0 and B are constants. Hence, the transistor acts as an **amplifier**, because a slight increase in emitter voltage produces a pronounced increase in collector current.

14.6.3 Field-Effect Transistor

The field-effect transistor (FET) employs a different transistor design. In the FET, a thin **channel** (hole conduction) is incorporated, located between the **source** and the **drain**, as shown in Figure 14.30.

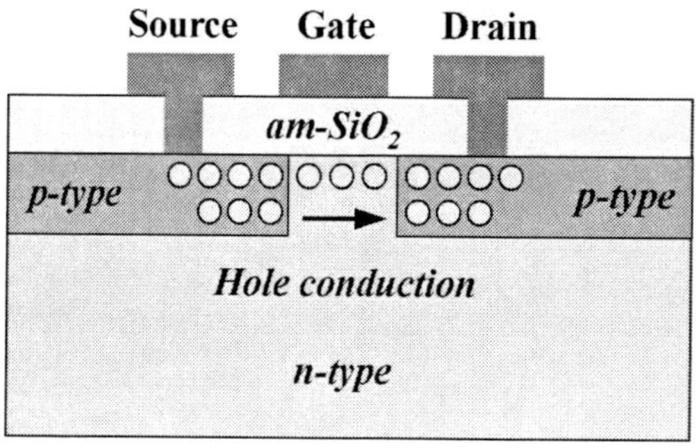

Fig. 14.30 Schematic of field-effect transistor with a thin channel between the source and the drain.

Note that the source and the drain correspond to the emitter and collector of a *npn*-transistor, respectively.

If a **negative voltage** is applied to the **gate** (which corresponds to the base in a *npn*-transistor), the **channel** *under* the *insulating layer*, composed of amorphous silica (am-SiO_2), becomes conductive. The field of the channel, which is a result of the *negative gate voltage*, produces an attraction for holes, because the semiconducting *n*-type material below the amorphous SiO_2 layer becomes distorted by the field to a *p*-type character. Therefore, holes can freely flow from the *p*-type source to the *p*-type drain. Altering or removing the external gate voltage affects the overall current.

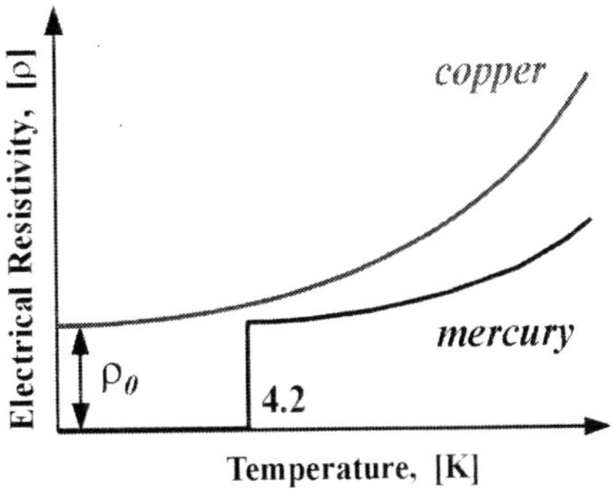

Fig. 14.31 Comparison between the electrical resistivity of copper and pure mercury. Note that the resistivity of mercury shows a sudden drop to zero at 4.2 K.

14.7 Superconductivity

As discussed earlier, the electrical resistivity of a metal such as silver or copper decreases steadily as temperature decreases and reaches a rather low residual value at 0 K.

In contrast, the electrical resistivity of pure **mercury** (Hg) reveals a sudden drop at 4.2 K, when temperature is continuously decreased (Figure 14.31). This particular behavior of mercury was discovered by **Heike Kammerlingh Onnes** [42] in 1911, earning him the Nobel Prize in 1913. It is interesting to note that this important result was a "byproduct" of Kammerlingh Onnes' research on liquefaction and solidification of helium.

The phenomenon of superconductivity is known for 26 metals (examples are Nb, V, and Pb) and more than 100 alloys and compounds. The effect was found to be **reversible** and is generally exhibited by metals that are rather poor conductors at room temperature. The drop in electrical resistivity at a characteristic temperature, which is termed **critical temperature**, T_c, is very sharp for pure metals, but may occur over the range of 1-2 K for metal alloys and compounds.

Until the 1980's, superconductor research focused especially on Nb alloys, but the transition temperature stayed below 25 K (Figure 14.32).

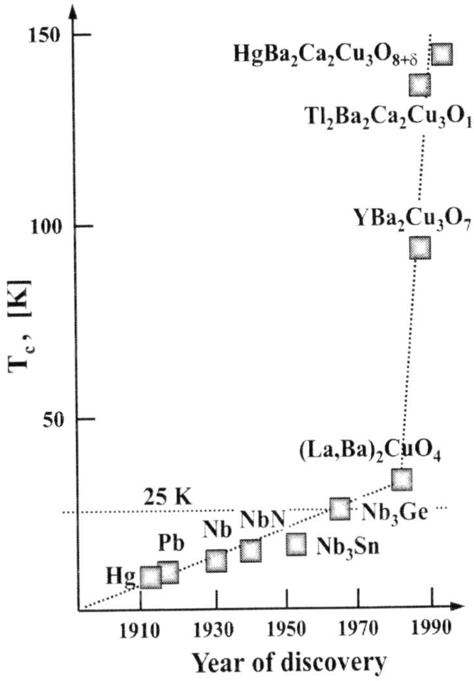

Fig. 14.32 Transition temperature, T_c, of various superconducting materials versus year of discovery.

---

[42] **Heike Kamerlingh Onnes** (1853-1926), Dutch physicist.

As Figure 14.32 shows, a dramatic increase in T_c began, when in April 1986, **G. Bednorz** and **A. Müller**[43] at the IBM laboratory in Zurich, Switzerland, reported exciting measurements of superconductivity in Ba-doped La_2CuO_4 at 35 K, the highest transition temperature at that time. Until 1986, the highest T_c was 23 K for Nb_3Ge thin films. In 1987, Bednorz and Müller received the Nobel Prize in physics for their revolutionary discovery.

In 1987, the ceramic **1-2-3 superconductor**, $YBa_2Cu_3O_{7-x}$ (1-2-3 refers to the three metal ion subscripts) was discovered with a T_c of 95 K. By 1988, a Tl-Ba-Ca-Cu-O-based ceramic compound was reported to have a transition temperature of 125 K and in 1993, the substitution of Hg for Tl in this material lead to a T_c of 133 K; this is the highest transition temperature to date. However, the toxicity of Hg and Tl strongly limits the application of such superconducting ceramic materials and hence the $YBa_2Cu_3O_{7-x}$ ceramic is still the most studied compound of such high-T_c cuperconductors.

The crystal structure (unit cell) of $YBa_2Cu_3O_{7-x}$ is rather complex, but closely related to the perovskite structure, $CaTiO_3$, with a triple unit cell of $YCuO_3$ perovskite, as shown in Figure 14.33.

If two of the three Y^{3+} ions are replaced by Ba^{2+} ions, the tetragonal $\mathbf{YBa_2Cu_3{}^{3+}O_8}$ structure is formed. *Charge neutrality* requires the removal of one oxygen ion per unit formula. This is achieved by the introduction of oxygen vacancies around the Y^{3+} ions.

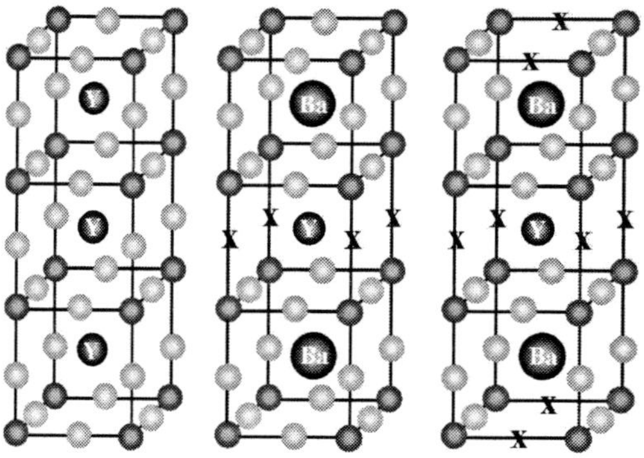

Fig. 14.33 Unit cell of $YBa_2Cu_3O_7$ derived from a triple $YCuO_3$ structure.

A characteristic feature of all cuprate superconductor structures is the presence of Cu in the divalent state, $\mathbf{Cu^{2+}}$. In Y-, Ba-, Cu-based superconductors, approximately 2/3 of the total Cu is divalent. Further incorporation of oxygen vacancies into the initial $Y_3Cu_3{}^{3+}O_9$ crystal structure leads to the formation of the $\mathbf{YBaCu_2{}^{2+}Cu^{3+}O_7}$ structure (compare Figure 14.33).

The oxygen stoichiometry, which varies proportionally with the valence state of Cu, is sensitive to temperature and oxygen partial pressure, and can vary between O_6 and O_7. The optimum oxygen stoichiometry for superconductivity is about $O_{6.92}$ (x=0.08). Since the **oxygen vacancies are ordered** along the [100]-direction, the tetragonal symmetry of $YBa_2Cu_3{}^{3+}O_8$ changes to an orthorhombic symmetry of the $\mathbf{YBaCu_2{}^{2+}Cu^{3+}O_{7-x}}$ structure.

Note that any isovalent substitution for Y^{3+} or Ba^{2+} commonly has only a moderate effect on the superconducting properties; however, substitution of other metals for Cu typically results in the loss of superconductivity.

[43] **Georg J. Bednorz** (1950-today), German physicist and **Alexander Karl Müller** (1927-today), Swiss physicist.

At sufficiently low temperatures, an ordering effect occurs between lattice atoms and electrons. This ordering effect is the **synchronization** between the vibrations of the atoms in the lattice and the wave-like motion of conducting electrons. At such low temperatures, the electrons are associated in pairs of opposite spin, termed **Cooper-pairs**, which are responsible for the complete loss of resistivity. The Cooper-pairs are named after **Leon N. Cooper** [44], who developed, together with John Bardeen and John R. Schrieffer, the theory of superconductivity (BCS-theory). In 1972, the Nobel Prize was awarded to Bardeen, Cooper, and Schrieffer for their jointly developed BCS-theory.

Note that in high-T_c superconductors, the copper-oxygen planes are the pathway for the super-current.

One important **limitation** of superconducting materials is the **critical current density**, J_c, which is defined as the maximum current flow at which the material is no longer superconducting. Unfortunately, the limitation in critical current density is more pronounced in high-T_c superconductors. This problem is caused by the penetration of a surrounding magnetic field into the material, creating an effective resistance due to the interaction between the current and mobile magnetic flux lines. Further research and material engineering is required to allow for a broad application of high-T_c superconducting materials.

Moreover, if a strong magnetic field is applied to a superconductor at temperatures below T_c, the superconductor will return to normal conduction. The magnetic field necessary to return to normal conduction is called the **critical magnetic field**, H_c, and is related to the critical temperature by:

$$H_c = H_0\left[1-\left(\frac{T}{T_c}\right)^2\right] \qquad (14.24)$$

where H_0 is the critical magnetic field at 0 K.

If a metallic or intermetallic superconductor is placed in a magnetic field, the magnetic field will penetrate through the superconducting material at temperatures above the critical temperature. However, if the temperature is below T_c and the applied magnetic field is below H_c, the magnetic field lines will be expelled from the material, with the exception of a very thin surface layer, as shown in Figure 14.34. This phenomenon of the exclusion of magnetic field lines (strong diamagnetic behavior; see also next chapter) in the superconducting state is called **Meissner effect** [45].

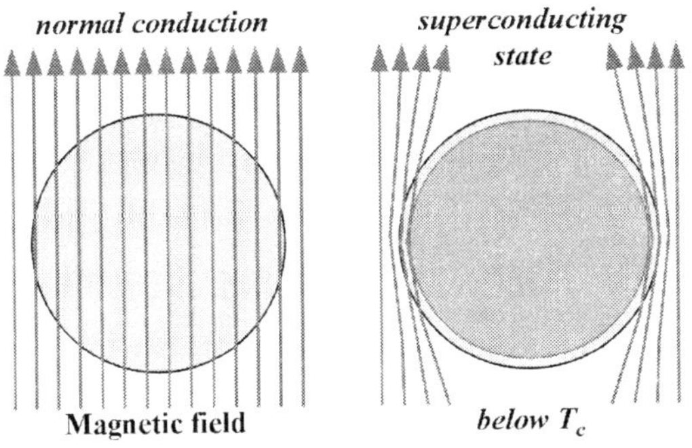

Fig. 14.34 Schematic illustrating the Meissner effect. Note that the temperature is below T_c and applied magnetic field is below H_c.

[44] **Leon N. Cooper** (1930-today), American physicist.

[45] **Walther Meissner** (1882-1974), German physicist.

15 Magnetic Properties

Magnetic materials are important for industrial applications, particularly in the area of electrical engineering. In general, there are two major types of magnetic materials:

- **hard** magnetic materials
- **soft** magnetic materials.

Soft magnetic materials are used in applications where easy magnetization and demagnetization is required. This group of materials is used, for example, as cores for power transformers and as stator and rotor materials for motors and generators. *Hard* magnetic materials, on the other hand, are required in applications where easy demagnetization should not occur. Examples include permanent magnets used in phone receivers, loudspeakers, or in particle accelerators.

15.1 Magnetic Field

The metals **Fe**, **Co**, and **Ni** are the only three elemental metals that can produce a strong magnetic field at room temperature, once they are magnetized. These metals are said to be **ferromagnetic**.

If a magnetized bar of Fe, Co, or Ni is placed below a sheet of paper and small iron particles are scattered on the paper surface, the particles will arrange along the **magnetic field lines** of the magnet. Such a bar magnet always has two poles (north and south) and the magnetic field lines appear to leave one pole and enter the other. In general, magnetism is **bipolar** in nature; no magnetic monopole has been discovered to date.

Note that this "magnetic dipole behavior" extends down to the atomic level.

A magnetic field can be produced by a current-carrying conductor, such as a long coil of Cu-wire. Such coils are termed **solenoids** and the **magnetic field strength**, H, is related to the number of turns, n, in the solenoid, the length, l, and the current, i, as given in equation 15.1:

$$H = \frac{0.4\pi \cdot n \cdot i}{l} \tag{15.1}$$

The magnetic field strength, H, has SI units of amperes per meter (A/m) and cgs units of oersteds (Oe). The conversion between SI and cgs units for H is: 1 A/m = $4\pi \times 10^{-3}$ Oe.

If a demagnetized iron bar is inserted into a long Cu coil (solenoid) and a magnetizing current is applied, the magnetic field outside the solenoid is then found to be stronger (Figure 15.1).

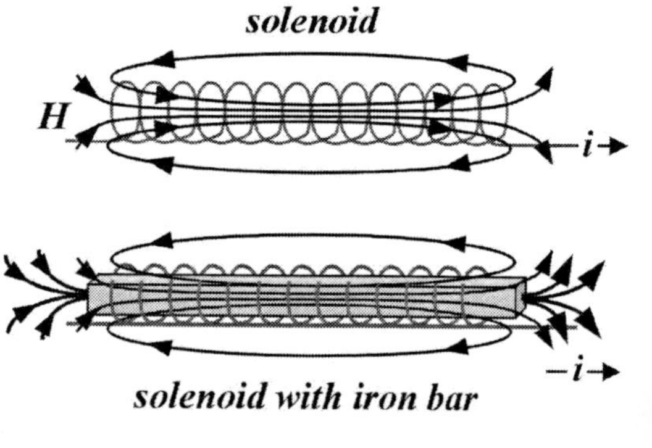

Fig. 15.1 Schematic illustrating the magnetic field, *H*, created around a solenoid with and without an iron bar. *Note* that the magnetic field outside the solenoid is stronger when an iron bar is inserted.

The enhanced magnetic field outside the solenoid is due to the sum of the applied field, H, and the external magnetic field of the magnetized iron bar placed inside the solenoid. This *new additive* magnetic field is called **magnetic induction,** B (or flux density, or simply induction). The **induced magnetic moment** (per unit volume) due to the insertion of the ferromagnetic iron bar is termed *intensity of magnetization* or **magnetization**, M, and is related to B and H via:

$$\begin{aligned} B &= \mu_0 H + \mu_0 M \\ &= \mu_0 (H + M) \end{aligned} \qquad (15.2)$$

where μ_0 is the permeability of free space (μ_0 = 4πx10[-7] tesla·meters per ampere (T·m/A)).

Note that μ_0 in equation 15.2 has no real physical meaning and is only used because SI units were chosen. The SI units for the magnetic induction B are webers (1 W = 1 V·s) per square meter or tesla [46]: 1 Tesla = 1 Wb/m² = 1 V·s/m² = 1.0·10⁴ G (gauss), and the SI units for H and M are amperes per meter (A/m).

Note that for many ferromagnetic materials the magnetization $\mu_0 M$ is much greater than the applied field $\mu_0 H$, which reduces equation 15.2 to:

$$B \approx \mu_0 M \qquad (15.3)$$

When a ferromagnetic material is placed in an applied magnetic field, the intensity of the magnetic field increases. This increase in magnetization is measured by a quantity called **magnetic permeability**, μ. The magnetic permeability is defined as the ratio of the magnetic induction B to the applied magnetic field H:

$$\mu = \frac{B}{H} \quad or \quad \mu_0 = \frac{B}{H} \qquad (15.4)$$

if there is only a vacuum within the applied magnetic field. The magnetic permeability of ferromagnetic materials is not a constant, but changes as the material is magnetized.

Note that the basic magnetic behavior is a *direct analog* of the electronic behavior. One can derive an alternative expression for Ohm's law (using $V = R \cdot I$ and $\sigma = l/R \cdot A$):

$$\frac{I}{A} = \sigma \cdot \frac{V}{l} \equiv B = \mu \cdot H \qquad (15.5)$$

where I/A is the current density and V/l is the voltage gradient. As can be seen from equation 15.5, the magnetic induction is analogous to the current density and the magnetic field strength is analogous to the voltage gradient, while the magnetic permeability corresponds to the electrical conductivity.

Furthermore, the magnetic permeability can be defined by the dimensionless quantity **relative magnetic permeability**, μ_r, which is a measure of the intensity of the induced magnetic field:

$$\mu_r = \frac{\mu}{\mu_0} \qquad (15.6)$$

and it follows:

$$B = \mu_0 \mu_r H \qquad (15.7)$$

Since the magnetization M of a magnetic material is proportional to the applied magnetic field, $M = \chi_m H$, the *proportionality factor* **magnetic susceptibility**, χ_m, is defined as:

$$\chi_m = \frac{M}{H} \qquad (15.8)$$

[46] **Nikola Tesla** (1856-1943), Yugoslavian-American inventor.

The magnetic susceptibility is similar to the relative magnetic permeability, μ_r, in that it is a dimensionless quantity and is often used when a material shows a weak magnetic response.

15.2 Types of Magnetism

Magnetic fields originate from the **movement** of the major electric charge carrier, the electron. When electrons move in a conducting wire, a magnetic field is produced around the wire, as shown in Figure 15.1 (solenoid).

The intrinsic magnetic properties of materials are also caused by the motion of electrons, but in this case, the magnetic field and the resulting magnetic force is a consequence of the intrinsic **electron spin** and the **orbital motion** of the electron about the nucleus (Figure 15.2).

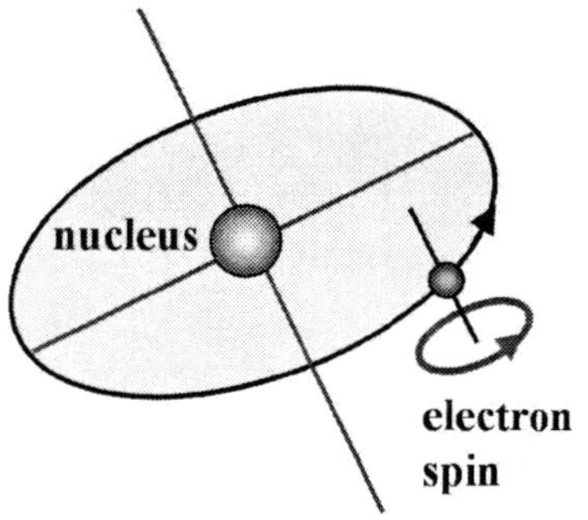

Fig. 15.2 Schematic showing the origin of magnetism (Bohr atom); the orbital motion of the electron about the nucleus and the electron spin.

15.2.1 Ferromagnetism

Ferromagnetism is a result of the **alignment of spins** of inner *unpaired electrons* in the crystal lattice. The most important ferromagnetic elements are Fe, Co, Ni, Gd (gadolinium), and rare-earth elements.

The **inner shells** of individual atoms are filled with *paired electrons* with opposing spins, and hence there is no resulting *magnetic dipole moment*. Similarly, the outer **valence electrons** are combined with each other due to chemical bonds. There is also no significant magnetic dipole moment caused by the outer valence electrons that participate in chemical bonding. Consequently, in most cases, electrons in atoms are paired and, thus, their negative and positive magnetic moments cancel.

However, unpaired electrons in inner electron shells can have a small positive dipole moment, as in the case for the 3d-electrons of Fe, Co, and Ni. Iron has four unpaired 3d-electrons (Hund's rule) and the spins of the 3d-electrons of adjacent atoms align parallel to one direction due to an interaction, which is referred to as **spontaneous magnetization**.

The parallel alignment of the magnetic dipoles of atoms is due to the creation of a **positive exchange energy** between individual dipoles (Figure 15.3).

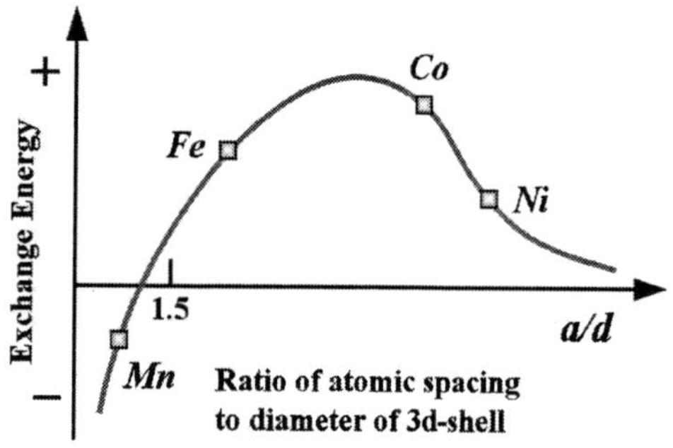

Fig. 15.3 Magnetic exchange energy versus the ratio of 3d-shell diameter to atomic spacing. *Note* that only elements with a positive exchange energy value are ferromagnetic.

The ratio of the atomic spacing to the diameter of the 3d-shell must be in the range of 1.4 to 2.7, in order to result in parallel dipole alignment. As can be seen from Figure 15.3, Fe, Co, and Ni are ferromagnetic, while Mn is not.

Each isolated electron that spins about its axis behaves as a single magnetic dipole. The corresponding magnetic dipole moment is termed **Bohr magneton**, μ_B, and can be expressed as follows:

$$\mu_B = \frac{eh}{4\pi m} \qquad (15.9)$$

where e is the electron charge ($1.6 \cdot 10^{-19}$ C), h is Planck's constant ($6.63 \cdot 10^{-34}$ J/s), and m is the electron mass ($9.11 \cdot 10^{-31}$ kg). Using equation 15.9, the value of one Bohr magneton can be calculated:

$$\mu_B = \frac{eh}{4\pi m} = \frac{(1.6x10^{-19}\,C)(6.63x10^{-34}\,Js^{-1})}{4\pi(9.11 \cdot 10^{-31}\,kg)}$$

$$= 9.27x10^{-24}\,A \cdot m^2 \qquad (15.10)$$

15.2.1.1 Temperature Effect

At any temperature above absolute zero (0 K), thermal energy causes the magnetic dipoles to deviate from their perfect parallel alignment. Hence, as temperature increases, the ferromagnetic effect gradually decreases until it completely disappears.

The temperature at which the ferro-magnetism is lost, is termed the Curie temperature, T_C, which is named after **Marie Curie**[47], who studied this phenomenon (Figure 15.4). As shown in Figure 15.4, the ferromagnetic behavior disappears at T_C and the material becomes paramagnetic. The Curie temperatures of Fe, Co, and Ni are 770, 1123, and 358 °C, respectively.

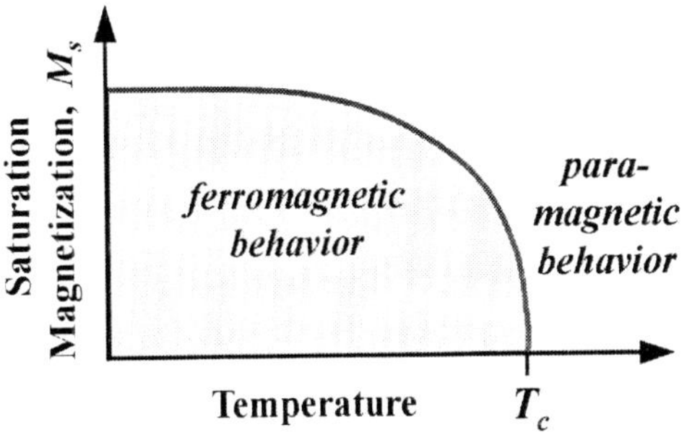

Fig. 15.4 Saturation magnetization, M_s, versus temperature. *Note* that at the Curie temperature, T_c, the material looses its ferromagnetic properties and becomes paramagnetic.

15.2.1.2 Ferromagnetic Domains

Below the Curie temperature, the magnetic dipole moments of ferromagnetic materials align parallel to each other. However, this process only occurs in microscopic regions within a single crystal, which are called **magnetic domains** (Figure 15.5).

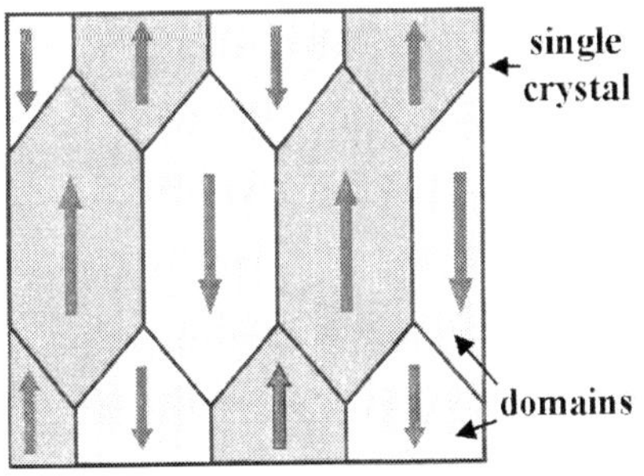

Fig. 15.5 Schematic showing random orientation of magnetic domains within a single crystal, resulting in zero net magnetization.

[47] **Marie Curie**, née Sklodowska (1867-1934), Polish physicist.

Note that when a ferromagnetic material is demagnetized by slowly cooling from above the Curie temperature, the individual magnetic domains will be randomly oriented, and no net magnetization of the crystal is recorded.

When an external magnetic field is applied to such a crystal, the magnetic domains, whose magnetic moments are initially aligned parallel to the applied field, will *grow* at the expense of those magnetic domains that are less favorably aligned (Figure 15.6).

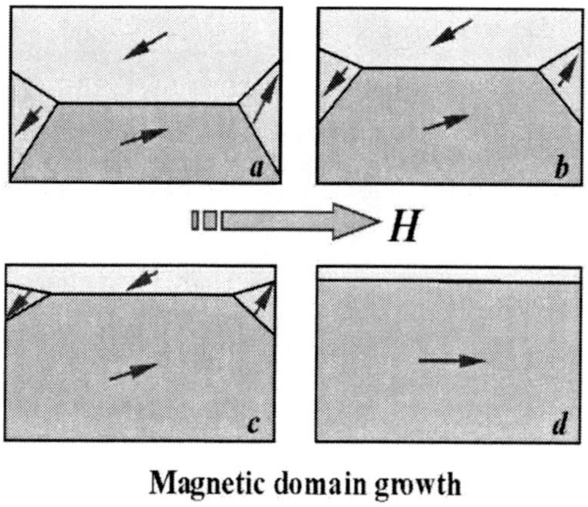

Fig. 15.6 Magnetic domain growth during magnetization of a ferromagnetic crystal.

Domain growth takes place via **domain wall motion** and the magnetization, M, increases as the applied magnetic field, H, increases. Domain growth can also occur via **domain rotation** (at high fields); however, growth due to domain-wall movement takes place *first*, since this process requires less energy.

When the applied magnetic field is removed, the magnetization of ferromagnetic materials *remains*, although a small amount of the magnetization is lost since some of the domains can rotate back to their original orientation.

The domain structure of a ferromagnetic material is stable, when the overall potential energy of the material is a minimum. The **total magnetic energy** of a ferromagnetic material is the sum of the following energy contributions:

- **exchange energy**
- **magnetostrictive energy**
- **domain wall energy**
- **magnetocrystalline anisotropy**
- **magnetostatic energy**.

15.2.1.3 Exchange Energy

The potential energy within a domain of a ferromagnetic solid is minimized when all its atomic dipoles are aligned in one direction. This alignment is associated with the **positive exchange energy** (see also Figure 15.3).

Note that even though the potential energy within a single domain is minimized, the external potential energy is increased by the formation of an external magnetic field.

15.2.1.4 Magnetostriction

When a ferromagnetic material is magnetized, its dimensions change slightly, and the sample either **expands** *or* **contracts** in the direction of magnetization. Iron, for example, shows a small expansion at low applied magnetic fields and a slight contraction at high fields.

The magnetically induced *reversible* elastic strain ($\Delta l / l$) is called **magnetostriction** and is on the order of 10^{-6}.

The cause of magnetostriction is attributed to the change in the bond length between atoms in ferromagnetic materials when their electron–spin dipole moments are *rotated* into alignment during magnetization. The fields of the dipoles may attract or repel each other, leading to the contraction or expansion of the crystal. The energy related to the mechanical stress created by the magnetostriction is termed **magnetostrictive energy**.

15.2.1.5 Domain Wall Energy

A domain wall is the boundary between two magnetic domains with different overall orientations of their magnetic dipole moments. It is similar to a grain boundary in a polycrystalline material, where the two adjacent single crystals reveal a different orientation. However, grain-boundary dimensions are typically on the order of only 1-2 nm (abrupt change in orientation), while domain walls are much wider. There is a gradual change in orientation of the magnetic dipole moments, as shown in Figure 15.7, which commonly extends to 100 nm in width.

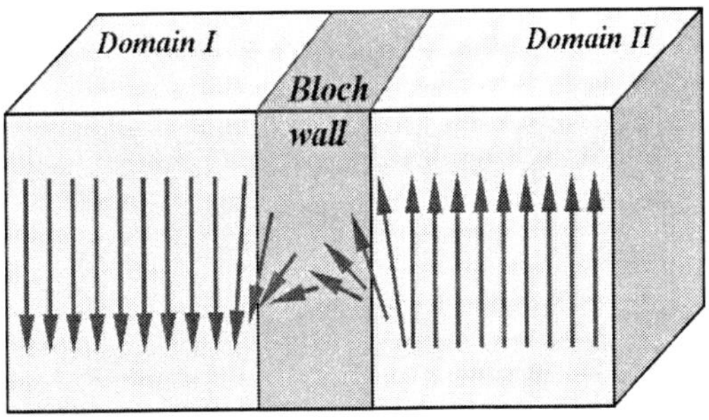

Fig. 15.7 A Bloch wall region between two domains with different orientation of their magnetic dipole moments.

This wide boundary region between two magnetic domains is also termed a **Bloch wall**. The reason for the large width of Bloch walls is due to a balance between two forces: (i) *exchange energy* and (ii) *magnetocrystalline anisotropy energy*, both of which determine the overall **domain wall energy** (Figure 15.8).

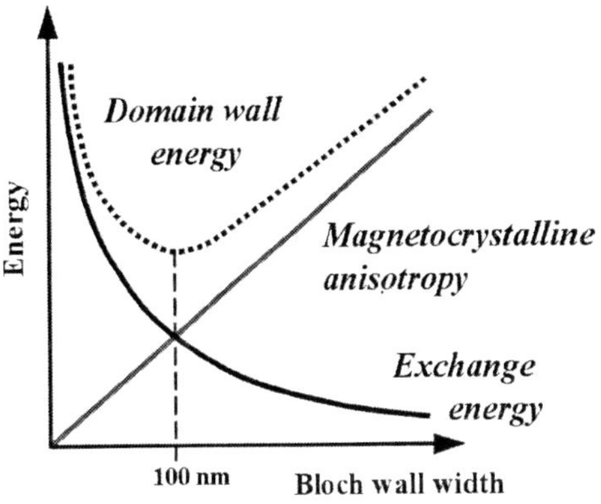

Fig. 15.8 Relation between exchange energy and magnetocrystalline anisotropy energy, which determines the overall domain wall energy (and Bloch wall width).

When there is only a small difference in orientation between magnetic dipoles, the exchange forces between the dipoles are minimized and the **exchange energy** is *reduced*. Thus, increasing exchange forces tend to widen the domain/Bloch wall.

However, the wider the domain wall dimension, the greater is the number of magnetic dipoles that are forced to lie in directions which are away from an orientation of easy magnetization. This in turn increases the **magnetocrystalline anisotropy energy**. Consequently, an equilibrium Bloch wall width is reached when the sum of the two energies is a minimum.

15.2.1.6 Magnetocrystalline Anisotropy Energy

In polycrystalline ferromagnetic materials, different grains with different orientation with respect to the applied magnetic field will reach **saturation magnetization**, M_s, at different field strengths. Grains with orientations close to the easy direction of magnetization will saturate at low fields, while grains away from a favorable orientation must rotate their magnetic dipole moments and, therefore, reach their saturation magnetization at much higher field strengths. The work done to rotate all the domains due to this orientation anisotropy is termed the **magneto-crystalline anisotropy energy**.

Note that in bcc-Fe single crystals, saturation magnetization occurs most easily in the <100> directions, while for the <111> directions a high applied magnetic field is required (hard directions). In contrast, for a fcc-Ni single crystal, the easy direction of magnetization are the <111> directions, while the <100> are the hard directions.

15.2.1.7 Magnetostatic Energy

The magnetostatic energy is the **potential magnetic energy** of a ferromagnetic material produced by its *external* magnetic field. This potential energy can be *minimized* by domain formation, as illustrated in Figure 15.9.

For a given unit volume of a ferromagnetic material, a single-domain structure has the highest potential energy; however, domain formation within the unit volume lowers the external field considerably. Since the intensity of the external magnetic field is directly related to its **magnetostatic energy**, the formation of multiple domains within a given unit volume lowers the magnetostatic energy.

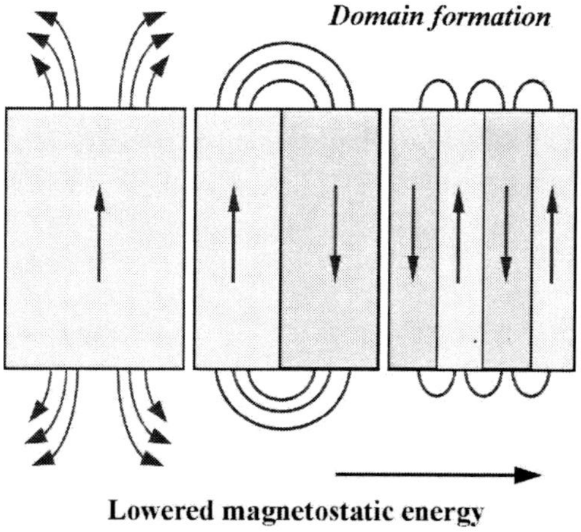

Fig. 15.9 Decrease in magnetostatic energy due to the formation of magnetic domains within the single crystal.

15.2.1.8 Hysteresis Loop

Ferromagnetic metals such as Fe, Co, and Ni acquire large magnetizations when placed in a magnetizing field. They remain magnetized, although to a lesser extent, when the external field is removed (Figure 15.10). When a magnetic field is applied to a demagnetized ferromagnetic metal (slowly cooled from above the Curie temperature), the magnetic induction, B, increases from zero as the applied field strength is increased. At one point, B reaches the **saturation induction**, B_s. Upon decreasing the applied field to zero, a magnetic flux density remains, which is termed **remanent induction**, B_r. To decrease the magnetic induction to zero, a reverse (negative) external magnetic field has to be applied, which is referred to as the

coercive force, H_c. If the applied negative field is further increased, the material reaches the *reverse saturation induction*, $B_{s(reverse)}$.

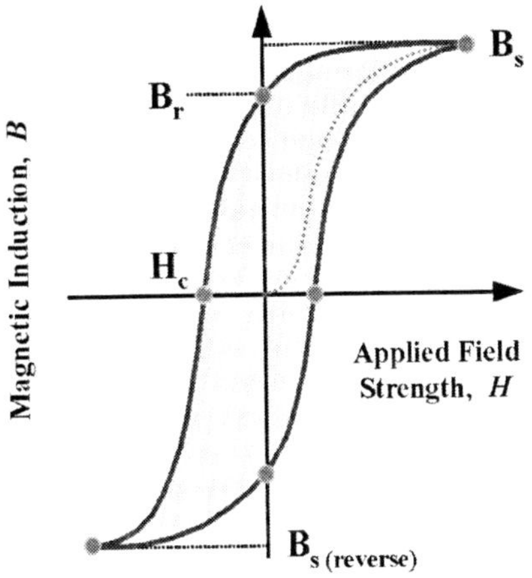

Fig. 15.10 Hysteresis loop of a ferromagnetic material (magnetic induction, *B*, versus magnetic field, *H*).

Upon removing the reverse field, the magnetic induction will return to the *remanent induction*. When a positive field is applied again, the B-H-curve will complete a loop. This magnetization loop is *reversible* and is referred to as the **hysteresis loop** of ferromagnetic materials.

15.2.1.9 Soft and Hard Magnets

The commercially important **metallic magnets** are *ferromagnetic*. In general, these materials can be either soft or hard magnets. A soft magnetic material is easily magnetized and demagnetized, whereas it is difficult to magnetize and demagnetize a hard magnetic material.

Soft magnets such as Fe-3%Si alloys are used in cores for transformers, motors, and generators and reveal a ***narrow*** **hysteresis loop**, which means their domain walls can easily be moved by an applied magnetic field (Figure 15.11).

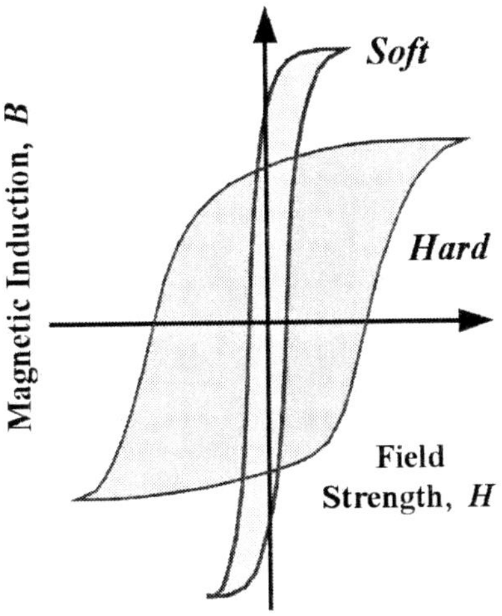

Fig. 15.11 Comparison of the hysteresis loops of soft versus hard ferromagnetic materials.

In contrast, **hard magnets** such as Cr-Co-Fe or Mn-Al-Fe alloys have a ***wide*** **hysteresis loop** with a high coercive force, H_c, and a high remanent magnetic induction, B_r, which means that a high external magnetic field has to be applied in order to demagnetize the material (see also the hysteresis loop in Figure 15.10) and that the domain walls are less mobile.

Note that the *area* of the hysteresis loop represents the *energy consumed* when traversing the loop. Hence, the small area of the hysteresis loop of soft magnets creates only a small energy loss when magnetization and demagnetization is repeatedly applied.

Another source of **energy loss** (in ac applications) is the generation of fluctuating electrical currents by the fluctuation of the magnetic field, which are termed **eddy currents**. The energy loss due to eddy currents is a result of **Joule heating** [48]. This energy loss can be reduced by increasing the resistivity of the material, which led to the development of higher-resistance Fe-Si-alloys in low-frequency power applications.

15.2.2 Diamagnetism

If an external magnetic field is applied to a material, the orbiting electrons of the atoms of the material will be slightly unbalanced. This creates small **magnetic dipoles** within the atoms, which *oppose* the external magnetic field.

This process can result in a **small *negative* magnetic effect** known as *diamagnetism*. As a result, a small negative magnetic susceptibility is typically found, which is on the order of $\chi_m = -10^{-6}$ (Table 15.1).

Note that diamagnetism occurs in all materials; however, in many materials, the small negative magnetic effect can be cancelled by a small positive effect. The diamagnetic effect remains only as long as the external magnetic field is maintained.

15.2.3 Paramagnetism

In contrast to diamagnetic materials (small negative effect), some materials show a **small *positive* magnetic susceptibility** in the presence of a magnetic field and are called *paramagnetic*.

Similar to diamagnetic behavior, the paramagnetic effect disappears when the applied magnetic field is removed. Paramagnetism produces magnetic susceptibilities ranging from 10^{-6} to 10^{-3} (see Table 15.1).

Tab. 15.1 Magnetic susceptibilities, χ_m, of some diamagnetic and paramagnetic materials.

Magnetic Susceptibility			
Dia-magnetics	χ_m x10^{-6}	**Para-magnetics**	χ_m x10^{-6}
Sn	-0.25	**O**	+106
Ag	-0.20	**Ti**	+1.25
Cd	-0.18	**Pt**	+1.10
Zn	-0.16	**Ca**	+1.10
Cu	-0.09	**Al**	+0.65

Note that paramagnetism is generated by the alignment of individual magnetic dipole moments of atoms or molecules due to an externally applied magnetic field. Since a high thermal energy input *randomizes* the orientation of magnetic dipoles, an increase in temperature decreases the paramagnetic effect.

The atoms of some transition and rare-earth elements possess incompletely filled inner shells with **unpaired electrons**. Apart from ferromagnetic behavior, these unpaired inner electrons can cause a *strong paramagnetic effect* since they are not counter-balanced by other bonding electrons.

15.2.4 Antiferromagnetism

In the presence of a magnetic field, magnetic dipoles of atoms can align in **opposing directions**. This phenomenon is called *antiferromagnetism* (Figure 15.12).

48 **James Prescott Joule** (1857-1894), English physicist.

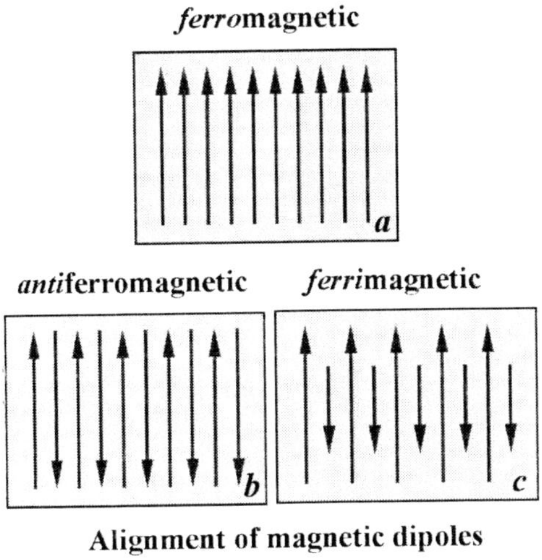

Fig. 15.12 Alignment of magnetic dipoles for three types of magnetic materials: (a) ferromagnetic, (b) antiferromagnetic, and (c) ferrimagnetic.

For example, the elements Mn and Cr exhibit *antiferromagnetism* at room temperature. They have a negative value of the exchange energy, because the ratio of their atomic spacing to the diameter of the 3d-orbit is less than 1.4 (see also Figure 15.3). Hence, **elemental compounds** that exhibit antiferromagentic behavior do not reveal a net magnetization, since their magnetic dipoles are oppositely aligned and are of the same magnitude.

15.2.5 Ferrimagnetism

Ferrimagnetism occurs in **compounds** with *multiple types* of cations such as in ceramic ferrites. The classical example is *magnetite*, Fe_3O_4, ($Fe^{2+}O{\cdot}Fe_2{}^{3+}O_3$), where divalent and trivalent Fe-ions are both present. When the magnetic moments of the ions with different valence are aligned *antiparallel*, similar to antiferromagnetic materials, there will be a net magnetic moment in one direction, since their magnetic moments do not cancel completely (Figure 15.12(c)).

In a **magnetite** crystal, the Fe^{2+} and Fe^{3+} ions are distributed corresponding to the **inverse spinel** structure. The magnetic moments of cations in octahedral and tetrahedral sites in the inverse spinel crystal structure are *antiparallel* (Figure 15.13).

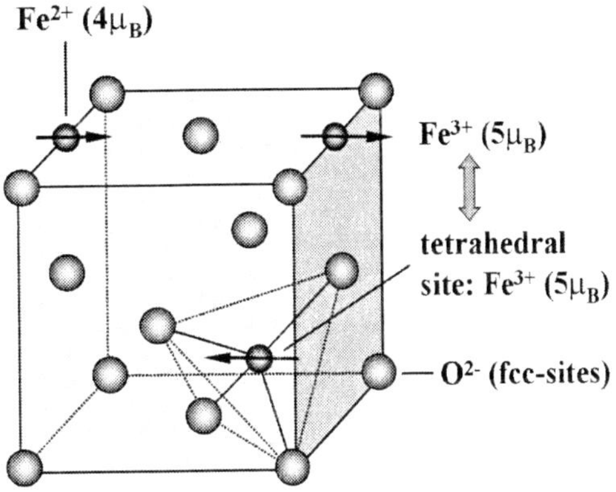

Fig. 15.13 Schematic showing the distribution of Fe^{2+} and Fe^{3+} between tetrahedral and octahedral sites in inverse spinels; only 1/8 of the unit cell is shown.

Therefore, an equal distribution of trivalent cations between these two sites leads to a *cancellation* of their magnetic moments. Consequently, the net magnetic moment of inverse spinels is only provided by the *divalent cations*. The number of **Bohr magnetons** provided depends on the divalent transition metals incorporated into the crystal lattice, as shown in Figure 15.14. The **net magnetic moment** of a unit cell of magnetite can be calculated using Figure 15.14.

Ion	3d-shell					#e⁻	
Fe^{3+}	↑	↑	↑	↑	↑	23	5
Mn^{2+}	↑	↑	↑	↑	↑	23	5
Fe^{2+}	↑↓	↑	↑	↑	↑	24	4
Co^{2+}	↑↓	↑↓	↑	↑	↑	25	3
Ni^{2+}	↑↓	↑↓	↑↓	↑	↑	26	2
Cu^{2+}	↑↓	↑↓	↑↓	↑↓	↑	27	1
Zn^{2+}	↑↓	↑↓	↑↓	↑↓	↑↓	28	0

Fig. 15.14 Number of electrons and Bohr magnetons of different transition metal ions.

There are eight tetrahedral sites (compare the one subcell shown in Figure 15.13) and sixteen octahedral sites, composing the inverse spinel unit cell. Since the magnetic moments of the trivalent cations cancel (opposite alignment of the dipole moments), only the eight Fe^{2+} ions with four Bohr magnetons each (occupying half of the octahedral sites of the *inverse spinel* structure) contribute to the net magnetization of the crystal:

$$8 \times 4\mu_B = 32\mu_B \qquad (15.11)$$

15.3 Ceramic Magnets

Most of the commercially important ceramic magnetic materials are **ferrimagnetic**. The dominant examples are ferrites, as discussed earlier, which are based on the *inverse spinel structure*.

Note that for *high-frequency* applications, no metallic alloy has a sufficiently high resistivity to prevent substantial *eddy-current losses*. Therefore, ceramic ferrites are widely used for such specific applications.

Although the term *ferrite* is often used to describe the group of magnetic ceramics, there are also other ceramic compounds that reveal interesting magnetic properties.

One example is the group of **garnets**, which have a relatively complex crystal structure, similar to the natural gem stone garnet, $Mg_3Al_2(SiO_4)_3$; *pyrope*. This crystal structure has *three* types of cation environments, as shown schematically in Figure 15.15.

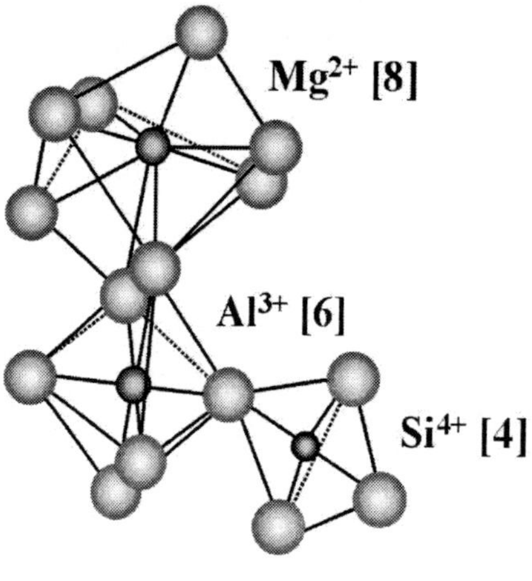

Fig. 15.15 Schematic illustrating the three cation environments (dodecahedral, octahedral, and tetrahedral) of the garnet structure, $Mg_3Al_2(SiO_4)_3$; pyrope.

In this structure, Mg^{2+} is in dodecahedral (8-fold) coordination, Al^{3+} is octahedrally coordinated, and Si^{4+} is in tetrahedral coordination. **Ferrimagnetic garnets**, such as the yttrium iron garnet (**YIG**), contain Fe^{3+}, which leads to the formula for YIG:

$Y_3^{3+}(Fe_2^{3+}, Fe_3^{3+})O_{12}^{2-}$. Here, the first two Fe^{3+} ions are in octahedral coordination, while the second three Fe^{3+} occupy tetrahedral sites. The three Y^{3+} ions are placed in the dodecahedral lattice sites. Magnetic garnets are often used as waveguide components in microwave communications.

Both ferrites and garnets are *soft* magnets, however, **hard magnetic ceramics** have also been developed. One commercially important example is the group of **magnetoplumbites**.

These materials have a hexagonal crystal structure and their chemical composition is based on the formula: $\mathbf{MO{\cdot}6Fe_2O_3}$, where *M* is a divalent cation, similar to the naturally occurring mineral *magneto-plumbite*, $Pb(Fe,Mn)_{12}O_{19}$.

This crystal structure is even more complex than the garnet structure with *five* different types of cation-coordination environments. The most commonly used divalent cations in magnetoplumbite ceramics are Sr^{2+}, Ba^{2+} and Pb^{2+}. These materials are used as **permanent magnets**, which are characterized by *high coercive fields* and *low processing costs*; the latter being a very important issue for introducing newly developed materials into the market.

Questions

What are the SI units for the magnetic induction, *B*?

What are the main types of magnetism?

What is the effect of magnetostriction on the bulk material?

What determines the width of Bloch walls?

Magnetite, Fe_3O_4, reveals ferrimagnetic properties; how is this magnetic behavior affected by the crystal structure?

Useful Equations

$$F_c = \frac{-K}{a^2} \quad (2.1)$$

$$F_R = \lambda e^{-a/\rho} \quad (2.3)$$

$$F_B = \frac{dE_B}{da} \quad or \quad E_B = \int F da \quad (2.5)$$

$$E = \frac{K_A}{a^6} + \frac{K_R}{a^{12}} \quad (2.7)$$

$$\cos\delta = \frac{\vec{D}\cdot\vec{D'}}{|D|\cdot|D'|} \quad (3.1)$$

$$= \frac{u\cdot u' + v\cdot v' + w\cdot w'}{\sqrt{u^2+v^2+w^2}\cdot\sqrt{(u')^2+(v')^2+(w')^2}}$$

$$d_{hkl} = \frac{a_0}{\sqrt{h^2+k^2+l^2}} \quad (3.3)$$

$$\sin^2\Theta = \frac{\lambda^2}{4a_0^{\ 2}}\left(h^2+k^2+l^2\right) \quad (3.4)$$

$$\ln(process\ rate) = \ln C - \frac{Q}{RT} \quad (4.2)$$

$$\frac{n_{defect}}{n_{lattice\ sites}} = C\cdot e^{-(E_{defect})/kT} \quad (4.5)$$

$$J_x = -D\frac{\partial c}{\partial x} \quad (4.9)$$

$$\frac{\partial c_x}{\partial t} = \frac{\partial}{\partial x}\left(D\cdot\frac{\partial c}{\partial x}\right) \quad (4.10)$$

$$\frac{c_x - c_0}{c_s - c_0} = 1 - erf\left(\frac{x}{2\sqrt{Dt}}\right) \quad (4.12)$$

$$\frac{c_s - c_x}{c_s - c_0} = erf\left(\frac{x}{2\sqrt{Dt}}\right) \quad (4.13)$$

$$D = D_0\cdot e^{-Q/RT} \quad (4.15)$$

$$\tau = G\cdot\gamma \quad (4.16)$$

$$\tau = \cos\phi\cdot\cos\varpi \quad (4.17)$$

$$F = C - P + 2 \quad (5.1)$$

$$\sigma = \varepsilon\cdot E \quad (6.4)$$

$$\varepsilon_{true} = \int_{L_0}^{L}\frac{dL}{L} = \ln\frac{L}{L_0} \quad (6.6)$$

$$\nu = -\varepsilon_x/\varepsilon_z \quad (6.7)$$

$$E = 2G(1+\nu) \quad (6.11)$$

$$\sigma_m = 2\sigma\left(\frac{c}{r}\right)^{1/2} \quad (6.13)$$

$$K_I = Y\sigma\sqrt{\pi a} \quad (6.14)$$

$$K_{IC} = Y\sigma_f\sqrt{\pi a_c} \quad (6.15)$$

$$t = C\exp(+Q/RT) \quad (6.18)$$

$$R = \frac{Md}{xmD} \quad (8.1)$$

$$y^2 = k_p t \quad or \quad y = k'_p t^{1/2} \quad (8.3)$$

$$MOR = \frac{3Fl}{2bh^2} \quad (9.1)$$

$$CaO \xrightarrow{ZrO_2} Ca''_{Zr} + O_O^x + V_O^{\bullet\bullet} \quad (9.2)$$

$$\tau = \frac{F}{A} = \eta\cdot\frac{d}{dx} \quad (9.4)$$

$$\Delta G(r) = 4\pi r^2\gamma_{sl} + \frac{4}{3}\pi r^3\Delta G_V \quad (9.5)$$

$$\overline{M}_m = \frac{\sum f_i M_i}{\sum f_i} \quad (10.1)$$

$$\sigma = \sigma_0\cdot e^{\frac{-t}{\tau}} \quad (10.5)$$

$$\frac{1}{\tau} = C \cdot e^{-Q/RT} \qquad (10.6)$$

$$E_c = E_f \frac{A_f}{A_c} + E_m \frac{A_m}{A_c} \qquad (11.8)$$

$$\frac{P_f}{P_c} = v_f \frac{E_f}{E_c} \qquad (11.12)$$

$$E_c = \frac{E_f E_m}{v_f E_m + v_m E_f} \qquad (11.21)$$

$$E_c^{\ n} = v_f E_f^{\ n} + v_m E_m^{\ n} \qquad (11.22)$$

$$C = \frac{Q}{\Delta T} \qquad (12.1)$$

$$c = \frac{q}{m\Delta T} \qquad (12.2)$$

$$\alpha = \frac{dL}{L \cdot dT} \qquad (12.4)$$

$$k = -\frac{dQ/dt}{A \cdot (dT/dx)} \qquad (12.5)$$

$$\Delta T_c = \frac{\sigma_f \cdot (1-\nu)}{E \cdot \alpha} \qquad (12.9)$$

$$\sigma_{surf} = \frac{E \cdot \alpha}{1-\nu} \cdot (T_a - T_s) \qquad (12.8)$$

$$E = h \cdot \nu = \frac{h \cdot c}{\lambda} \qquad (13.1)$$

$$\frac{n}{n'} = \frac{\sin \Theta_r}{\sin \Theta_i} \qquad (13.3)$$

$$R = \left(\frac{n-1}{n+1} \right)^2 \qquad (13.5)$$

$$\frac{I}{I_0} = e^{-\alpha \cdot t} \qquad (13.6)$$

$$(1-R)(I_o \cdot e^{-\alpha \cdot t}) \qquad (13.7)$$

$$(R)(1-R)(I_o \cdot e^{-\alpha \cdot t}) \qquad (13.8)$$

$$I_t = (1-R)^2 (I_0 \cdot e^{-\alpha \cdot t}) \qquad (13.9)$$

$$\lambda = h \cdot c / E_g \qquad (13.10)$$

$$\ln \frac{I_l}{I_0} = -\frac{t}{\tau} \qquad (13.11)$$

$$V = I \cdot R \qquad (14.1)$$

$$\rho = \frac{R \cdot A}{l} \qquad (14.2)$$

$$R = \frac{\rho \cdot l}{A} \qquad (14.3)$$

$$\sigma = \frac{1}{\rho} \qquad (14.4)$$

$$\sigma = n \cdot q \cdot \mu \qquad (14.5)$$

$$\sigma = n_n q_n \mu_n + n_p q_p \mu_p \qquad (14.7)$$

$$\rho = \rho_{0^\circ C} (1 + \alpha \cdot T) \qquad (14.8)$$

$$f(E) = \frac{1}{e^{(E-E_f)/kT} + 1} \qquad (14.9)$$

$$\sigma = \sigma_0 \cdot e^{-E_g/2kT} \qquad (14.15)$$

$$\ln \sigma = \ln \sigma_0 - \frac{E_g}{2kT} \qquad (14.16)$$

$$n_p = \frac{n_i^{\ 2}}{p_p} \approx \frac{n_i^{\ 2}}{N_a} \qquad (14.21)$$

$$I_c = I_0 e^{V_e/B} \qquad (14.23)$$

$$B = \mu_0 (H + M) \qquad (15.2)$$

$$\mu = \frac{B}{H} \quad or \quad \mu_0 = \frac{B}{H} \qquad (15.4)$$

$$B = \mu_0 \mu_r H \qquad (15.7)$$

$$\chi_m = \frac{M}{H} \qquad (15.8)$$

$$\mu_B = \frac{eh}{4\pi m} \qquad (15.9)$$

Index